THE INTERNATIONAL FIRE SERVICE TRAINING ASSOCIATION

The International Fire Service Training Association is an educational alliance organized to develop training material for the fire service. The annual meeting of its membership consists of a workshop conference which has several objectives —

. . . to develop training material for publication
. . . to validate training material for publication
. . . to check proposed rough drafts for errors
. . . to add new techniques and developments
. . . to delete obsolete and outmoded methods
. . . to upgrade the fire service through training

This training association was formed in November 1934, when the Western Actuarial Bureau sponsored a conference in Kansas City, Missouri, to determine how all agencies that were interested in publishing fire service training material could coordinate their efforts. Four states were represented at this conference and it was decided that, since the representatives from Oklahoma had done some pioneering in fire training manual development, other interested states should join forces with them. This merger made it possible to develop nationally recognized training material which was broader in scope than material published by an individual state agency. This merger further made possible a reduction in publication costs, since it enabled each state to benefit from the economy of relatively large printing orders. These savings would not be possible if each individual state developed and published its own training material.

From the original four states, the adoption list has grown to forty-four American States; six Canadian Provinces; the British Territory of Bermuda; the Australian State of Queensland; the International Civil Aviation Organization Training Centre in Beirut, Lebanon; the Department of National Defence of Canada; the Department of the Army of the United States; the Department of the Navy of the United States; the United States Air Force; the United States Bureau of Indian Affairs; the United States General Services Administration, and the National Aeronautics and Space Administration (NASA). Representatives from the various adopting agencies serve as a voluntary group of individuals who govern policies, recommend procedures, and validate material before it is published. Most of the representatives are members of other international fire protection organizations and this meeting brings together men from several related and allied fields, such as:

. . . key fire department executives and drillmasters,
. . . educators from colleges and universities,
. . . representatives from governmental agencies,
. . . delegates of firefighter associations and organizations, and
. . . engineers from the fire insurance industry.

This unique feature provides a close relationship between the International Fire Service Training Association and other fire protection agencies, which helps to correlate the efforts of all concerned.

The publications of the International Fire Service Training Association are compatible with the National Fire Protection Association's Standard 1001, "Fire Fighter Professional Qualifications (1974)," and the International Association of Fire Fighters/International Association of Fire Chiefs "National Apprenticeship and Training Standards for the Fire Fighter." The standards are an effort to attain professional status through progressive training. The NFPA and IAFF/IAFC Standards were prepared in cooperation with the Joint Council of National Fire Service Organizations of which IFSTA is a member.

In order to formally adopt IFSTA training material, the firefighters should first request it through their training agencies. While adopting does not obligate any sponsoring agency, it still profits by sending a representative to the Annual IFSTA Validation Conference. This representation permits a voice and a vote on the ever-changing policies, techniques, and procedures which must be reflected in each publication. Adoption further adds prestige to any training program and this, in turn, adds prestige to IFSTA publications.

The International Fire Service Training Association meets each July at Oklahoma State University, Stillwater, Oklahoma. Fire Protection Publications at Oklahoma State University publishes all IFSTA training manuals and texts. This department is responsible to the executive board of the association. While most of the IFSTA training manuals can be used for self-instruction, they are best suited to group work under a qualified instructor.

Preface

Since 1937, IFSTA has developed and published training manuals for the fire service. The greater part of those training manuals have developed the manipulative skills that are so essential to the firefighter's performance of his duties. By having access to this body of knowledge, the fire service has been able to greatly improve its ability to cope with the fire problem.

AMERICA BURNING was published in 1972 as a Report of the President's Commission on Fire Prevention and Control. This committee report identified the number one fire problem in America as public apathy and ignorance. The fire service must share the greater part of the responsibility for the public attitude toward the fire problem. If we are to reduce the incidence and the severity of fire, it must be achieved by methods other than the conventional attack by fire suppression forces. This manual was developed to give the firefighter some basic knowledge concerning the problem of reaching the public with an educational program. For those who have a genuine concern for the lives and the safety from fire of the people who are entrusted to their care, we dedicate this volume.

Tribute must be paid to the committee for the method and manner by which this manual was produced. Many members, experts in their fields, wrote individual chapters and then worked with the group to mold the material into a single continuous publication. We owe a great debt to all of those who have participated in the development of this manual. They have been and are willing to share their knowledge that others may be better qualified to teach the public about fire.

Those who have contributed chapters to this work are:

Becky Carnes, R. N.	Nancy Newman, R. N.
Donald W. Davis	Dennis Ozment
Lonnie Jackson	Lawrence Pairitz
Debby Landis, R. N.	Roger E. Salisbury, M.D.
Cathy Lohr	Donald Shields, Ph.D
Thomas Makey	Dick Small
Helen Moskal	Jim Smalley

Nancy Trench

IFSTA 606   FIRST EDITION

PUBLIC FIRE EDUCATION

WRITTEN BY A SPECIAL IFSTA COMMITTEE OF PUBLIC FIRE EDUCATORS (see page iv)

VALIDATED BY

THE INTERNATIONAL FIRE SERVICE TRAINING ASSOCIATION

PUBLISHED BY

FIRE PROTECTION PUBLICATIONS ● OKLAHOMA STATE UNIVERSITY

Jerry W. Laughlin - Editor
Gene P. Carlson - Associate Editor
Connie E. Osterhout - Assistant Editor

PUBLISHED BY **Fire Protection Publications Oklahoma State University**

In cooperation with
The International Association of Fire Fighters
The Insurance Services Office
Numerous State Boards for Vocational Education

Dear Firefighter:

The International Fire Service Training Association (IFSTA) is a nonprofit organization that exists for the sole purpose of serving firefighters. IFSTA is a member of the Joint Council of National Fire Organizations, National Fire Protection Association and International Society of Fire Service Instructors. If you need help in locating additional information concerning the organization, training materials, or manual orders, please contact one of the people listed below.

Without you we do not exist!

Respectfully,

Jerry W. Laughlin

Write:
Customer Services
Fire Protection Publications
IFSTA Headquarters
Oklahoma State University
Stillwater, Oklahoma 74074

or call:
(405) 624-5723
Manual Orders — Brenda Ingersol

Oklahoma State University in compliance with Title VI of the Civil Rights Act of 1964 and Title IX of the Educational Amendments of 1972 (Higher Education Act) does not discriminate on the basis of race, color, national origin or sex in any of its policies, practices, or procedures. This provision includes but is not limited to admissions, employment, financial aid, and educational services.

©1979 by the Board of Regents, Oklahoma State University
All rights reserved
ISBN 0-87939-034-4
Library of Congress 79-89165
First Edition published 1979
Printed in the United States of America

Those who worked on the committee in editing, suggesting and adding their expertise are:

Cathy Lohr

Howard Boyd   Garland Fulbright
Larry Borgelt   Ed McCormack, Jr.

Elton Nixon

Our gratitude is also extended to the following individuals on the IFSTA staff who made the final publication of this manual possible:

C. Ann Moffat - Artist   Carol Smith - Proofreader
E. T. Brown, III - Photographer   Nancy Wilson - Phototypesetter Operator

Other acknowledgements go to Pam Powell of the United States Fire Administration, who edits their Resource Exchange Bulletin, and to the Burger King Corporation, the Stillwater (Oklahoma) Fire Services, Gene Duncan, Carolyn Thomas and at least a dozen others.

A number of other individuals and companies were very generous with their time and materials during the production of this book. Unfortunately it is never possible to be sure of listing everyone who contributed something, and that is regrettable because the editorial staff is dependent upon the assistance of many helpful people each time a book is published. Certainly every contribution is appreciated, and it is hoped that each individual involved, whether listed or not, can take pride in what has resulted.

IFSTA 606 — Public Fire Education
          Editorial Review Committee

Chairman — Howard Boyd, Nashville (Tennessee) Fire Department
Vice-Chairman — Lonnie Jackson, Mount Prospect (Illinois) Fire Department
Secretary — Cathy Lohr, North Carolina Department of Insurance

Larry Borgelt • Donald W. Davis • Garland W. Fulbright • Edward H. McCormack, Jr. •
Elton Nixon • Dennis Ozment • James C. Smalley • (See back of book for affiliations)

Contents

	INTRODUCTION	**1**
1	**PUBLIC FIRE EDUCATION PLANNING**	**9**
	The Process	11
	The People	12
	Starting Point: Establish Responsibility	12
	Step One: Identification	14
	Step Two: Selection	18
	Step Three: Design	23
	Step Four: Implementation	26
	Step Five: Evaluation	30
	Worksheets for Planning	32
2	**TARGET AGE GROUPS**	**35**
	Fire Education Learning Objectives for Preschool Children	36
	Stop, Drop and Roll!	36
	Matches	40
	Exit Drills in the Home (EDITH)	42
	Hazard Recognition	45
	Public Fire Education Planning — Grades 1 - 3	46
	Public Fire Education Planning — Grades 4 - 6	50
	Fire Education Planning — Junior and Senior High School	53
	The Youthful Fire-Setter	54
	Adult Fire Education	54
	Fire Education for Older Citizens	57
	General Suggestions	58
3	**SEASONAL AND SPECIFIC GROUP PROGRAMS**	**59**
	Posters and Signs	60
	Fire Prevention Week	62
	Home Fires	66
	Proper Means of Reporting a Fire	70
	Mobile Homes	70
	Apartments	71
	Volunteer Programs	72
	Fire Station Tours	73
	Christmas	74
	Halloween	76

	July 4th	76
	Thanksgiving	76
	Spring Clean Up	76
	Nursing Homes	77
	Churches	78
	Fire Safety in Recreational Environments	78
	Grass and Forest Fire Hazards	82
4	**SMOKE DETECTORS**	**83**
	Two Types of Smoke Detectors	84
	Power Sources	86
	Select Location	88
	Interconnecting Smoke Detectors	91
	Testing	91
	Maintenance	92
	Heat Detectors	93
	Key Factors for Escape	94
	Act Now	95
	Suggested Lesson Plan for Fire Education Program	96
	Detectors Not Enough	101
	Review	102
	Summary Statement	103
5	**WORKING WITH THE MEDIA**	**105**
	Match the Media and the Message	106
	General Tips for Working with the Media	108
	Acknowledge Assistance	111
	Samples of Media Messages	111
6	**USING VISUALS**	**115**
	Slide Programs	116
	Five-Screen Slide Program	120
	Table Top Demonstrations	122
	Aerosol Flammability	124
	Candle and Jar	124
	Candle Box	124
	Candle Box CO_2	125
	Illuminated Fire Triangle	126
	Flameproofing	127
	Extinguishment with Lid	127
	Baking Soda Extinguishment	127
	Smokehouse	127
	Dust Explosions	128
	Vapor Trail Tube	128
	Vapor Explosion Tube	129
	Demonstration Kits	130
	Demonstration Example	132
	Electrical Cords - Flannel Board Story	134
	The Story of the Little Red Fire Hat	136

7   RESOURCE EXCHANGE   141

What Is a Resource?   142
What Is Exchange?   143
Resource Exchange Systems   143
Components of the System   143
System Operations   145
System Complexity   147
The People Involved In Resource
    Exchange   148
The Information Processor/Systems
    Manager   149
The Contributor   150
The Receiver   150
Summary   150

APPENDIX A: Blueprints for Planning   152

APPENDIX B: Burn Injuries   161

Dedication

This manual is dedicated to the members of that unselfish organization of men and women who hold devotion to duty above personal risk, who count sincerity of service above personal comfort and convenience, who strive unceasingly to find better ways of protecting the lives, homes and property of their fellow citizens from the ravages of fire and other disasters . . . **The firefighters of All Nations.**

Introduction

The fire service is the oldest protection service in the United States. Roots go back to New Amsterdam in 1635 when the Dutch settlers established a "rattle watch" to patrol the streets to be alert to fire and other happenings.

The thatched roofs and wood construction were constantly a hazard. Each citizen was required to keep buckets which would be available when the cry "Throw out your buckets" was given. They were to respond by forming a bucket brigade to help fight fires.

The fire service eventually progressed from buckets and untrained citizen brigades to manually operated pumps and organized volunteer fire companies. Many famous men were among the ranks of those early firefighters, including George Washington, who purchased a pumper for his fire company from his own funds. Volunteer firefighters, then as today, came from all walks of life to contribute their skills to the saving of lives and property in their community.

The only fire protection before the Civil War was provided by volunteer companies. Large cities turned to paid departments for fire protection. The fire service later progressed to horse-drawn equipment, to mechanized gasoline engines, and finally to our present Diesel-driven pumpers and the best fire fighting equipment available to man. As society progressed, fire departments, likewise, advanced their fire protection capabilities.

Fire fighting techniques, apparatus and equipment have greatly advanced from the horse-drawn fire engine days. Efforts in the past have been directed mainly toward advancing fire suppression methods with less emphasis on fire prevention. The modern trend is, however, toward directing more attention to prevention and public education in order to reduce fire incidents.

The complexities of fire prevention and control increased dramatically with the construction of larger, more closely spaced buildings and ever-increasing population which, even today, does not understand the constant danger fire presents to them.

There is an old saying in the fire service: *"The three main causes of fires are MEN, WOMEN and CHILDREN."* One look at all the commonly listed causes of fire indicates that human carelessness is, indeed, to blame. Children playing with matches, the careless use of smoking materials and flammable liquids are only a few examples of potentially dangerous human behavior. Examine all fire causes and you will find a wrong human behavior of some sort, either accidental or intentional, behind most fires.

The average person when involved in a fire will behave in certain predictable ways, such as opening a hot door to escape, or running while his clothing is on fire.

Upon close examination of the fire problem it may be observed that the cause of most fires is the result of the same mistakes, just being made by different people at different times. Fire is not a forgiving adversary and many victims are severely injured or killed before they can correct their actions. Annually we are burning more of our human and economic resources than any other major nation on our planet.

There have been many fires which killed large numbers of people at one time, such as the Cocoanut Grove night club fire (492) and the Iroquois Theater fire (602). Even as recent as 1977, fire killed over 160 in a single incident at a night club/restaurant fire in Kentucky. These and other disasters resulted in the establishment of building and fire prevention codes that have been highly effective in reducing subsequent fire incidents and losses. Many other codes have come into existence because of known hazards learned through experience. Still, fires are destroying buildings, killing citizens and firefighters because someone, out of apathy, ignorance or both, allowed dangerous conditions to exist.

Tragedies such as the Beverly Hills Supper Club fire in Kentucky have motivated individuals to enact and enforce stricter building and fire prevention codes. These incidents are indeed tragic and require action to prevent future occurrences. Fires in homes, apartments and other dwellings claim many more lives a year and yet little attention has been directed toward fire prevention in this area.

The American home, whether it be a single-family dwelling, mobile home, or high-rise apartment, can be a very dangerous place, especially when it is abused by its occupants. Normally, buildings are constructed within adopted codes and many public buildings are inspected periodically, but little consideration is given to the private home.

Effective code enforcement, public fire education in school systems and civic groups, news media promotion and many other innovative sources can offer various ways to attack the fire problem. The objective is to teach people what they can do to change their life styles to keep fires from happening; how to select and use smoke detectors to give those precious few minute's warning when prevention methods fail; how to survive in smoke and heat through exit drills in the home; and how to get out, notify the fire department and live.

The number of fires will be reduced through safer living habits; less fire loss can be obtained by a faster call to the fire department by people awakened by a smoke detector; and peoples' lives can be saved by their getting out faster. Burns and scalds can also be reduced through the same programs by targeting this problem and using effective programs to help in the area of electrical burns, boiling liquids on stove tops, chemicals that burn and hot tap water.

Today's firefighter must become a communicator of facts relating to the causes and hazards of fire, a subject he knows well because he has seen it often. Just as the firefighter is trained in the nuts and bolts aspects of his job, he also needs to be trained in small group communications to effectively deal with the public.

Skilled communication includes knowing how to listen and talk; how to see that the idea is being received and transmitted by another. Skilled communications techniques are as important as the message itself. These skills are developed through training, refinement and perseverance. Such training should be at the core of an effective fire safety education commitment.

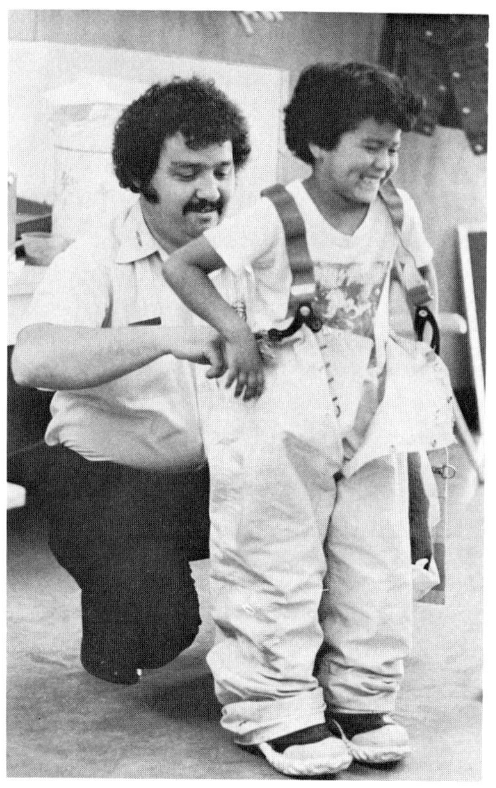

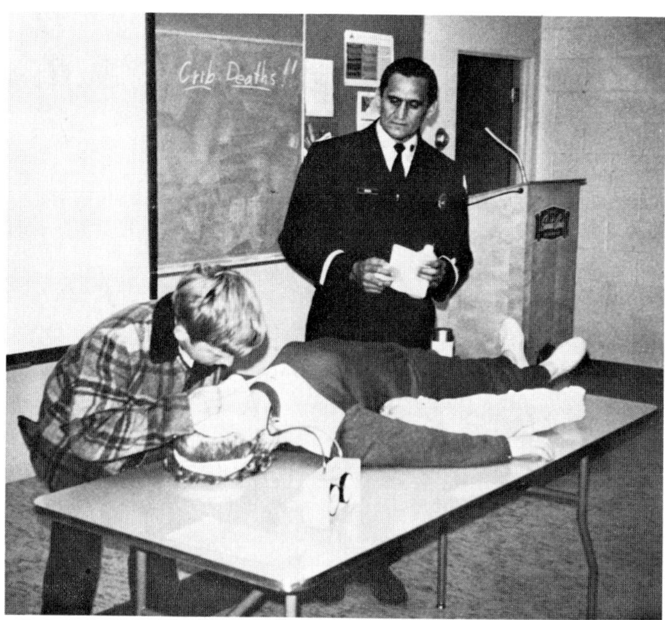

Designing a public fire education program requires contributions of both skill and knowledge from individuals. Once the program is designed, it is important that the people selected to present the program have good communications skills.

Community groups and associations play increasingly important roles in support of the fire service. Such groups assist by initiating and supporting drives to stimulate greater citizen interest and provide support for fire service personnel. Fire services in many areas have learned that such groups can spearhead drives to raise funds for special equipment; provide scholarships and death funds for firefighters' dependents; and, in general, assist in bridging the gap between the fire service and the people in communities. No doubt these groups are effective influences for increased and improved services. What is less understood is that fire service/community group relations can intertwine to service the broader function of the community. The plans and decisions that can result from actively sought, well conceived community involvement contribute to long-term benefits. Programs developed through such fire service/community interface would be the result of both citizens' ideas and the professional firefighter public education staff. The success of such cooperative efforts will benefit the fire service and the community. Open participation on such fire education projects as sprinkler laws, smoke detectors, home exit drills, brigade training and even community betterment projects will show the fire service's concern for the community and, in turn, the community's concern for the fire service.

Community involvement is essential to successful public fire education programs

There are several methods to involve the public and the fire service in joint projects that will enhance the spirit of civic involvement of all participants. For example, the fire service can work in team efforts with the community, such as a fire safety seminar that proves useful to both the firefighters and the community. The fire service can also plan, develop and implement a series of public information programs staffed by firefighter/citizen teams. Finally the fire service might use the mass media (television, radio, press) to dramatize achievements, problems and changes that relate to both the community and the service.

Over a period of time, routine daily contacts coupled with specific planned efforts can do much to influence the overall impression which members of the public, city officials and community councils have of the fire service.

Can Public Education reduce deaths and injuries? A cynic might remark that widespread ignorance of fire proves that Fire Prevention Week, school programs in fire safety and all the posters and pamphlets on fire prevention are wasted efforts. Yet, we do not know how much worse the nation's fire record would be if there were no educational efforts. On the other hand, we do know that public education programs can dramatically reduce fire losses. Two studies supported by the Bureau of Community Environmental Management, an arm of the Department of Health,

Public fire education programs CAN reduce fire losses

Education and Welfare, provide evidence of this. Though small in scope, the studies are among the few in which results of fire prevention efforts have been measured.

Between 1966 and 1969, an intensive fire safety education program was directed at an area of southeast Missouri where the fire death rate was far higher than the national average. The first step was to study the pattern of fires and burn injuries and their causes. Then a field staff was trained to administer the program. Civic groups, fire departments, local officials and the mass media cooperated in the program. The public received fire safety messages every way they turned — audio-visual demonstrations, educational programs and media broadcasts. The result: The fire death rate dropped 43 percent in three years — from 12.9 to 7.4 per 100,000 population. For each dollar invested in the program, 20 dollars were saved in anticipated property losses, medical expenses and earning losses. Two years after the pilot program was terminated, the fire death rate was still falling — five times faster than that of the rest of the state.

A similar study had been carried out eight years earlier in Mississippi County, Arkansas. Their studies showed that misuse of electrical wiring systems and petroleum products, plus use and storage of flammable products near heating units, led other causes of fire. The public education program emphasized these problems. Following the first year of the education program there were only half as many burn injuries requiring medical treatment as the year before. This favorable trend continued during ensuing years.

A number of incidents in recent years has demonstrated that when people have fire safety on their minds, fires decrease in number. In each incident, people were fire-conscious because they knew normal fire protection was not available to them. It happened in a midwestern city when a severe snowstorm immobilized all traffic, including fire apparatus. It happened in several American cities in the late 1960's when fire departments were tied up in riot-torn areas. It has happened when fire departments have been battling landslides or coping with floods. In each case, the number of fires dwindled to a fraction of the normal.

A striking example of long-term success in fire safety education is the Smokey the Bear campaign. That effort, supported by Federal and State forest agencies, has been described as the country's most successful program of environmental protection. For 30 years public service advertising has urged Americans to prevent forest fires. During these years, man-caused forest fires have been reduced from about 200,000 annually to about 105,000 in 1971. This reduction was achieved in spite of a

The Smokey the Bear campaign has successfully reduced man-caused forest fires. This is only one example where public fire education can indeed reduce the losses rendered by fire.

doubling in land area covered by the statistics and a tenfold increase in recreational use of the land. A doubling of the acreage alone would be expected to result in 400,000 fires annually, but, as indicated, only 105,000 occurred. This overall reduction by 75 percent in the number of fires which would otherwise be expected to occur (assuming that the increased exposure to people leads in equal measure to chance of fire and the chance of early detection) has helped save $17 billion in natural resources over the 30-year period. The cost of this program to Federal and State agencies is about $488,000 per year, with approximately $40 million in service donated by the nation's radio and television stations, newspapers, magazines and the Advertising Council.

FIREMAN
FRED
You're
Big Enoug
for
Fire Safe

1 *Public Fire Education Planning*

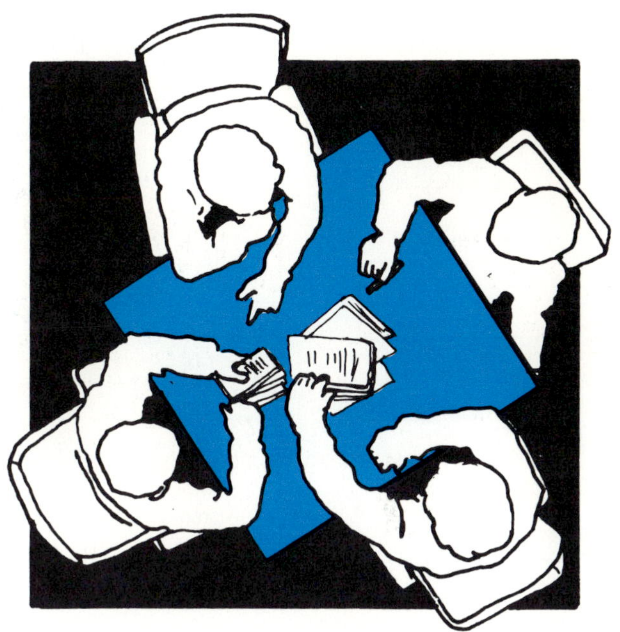

Public fire education planning describes a systematic approach to designing, implementing and evaluating those programs conducted to instruct the public about fire safety.

This chapter consists of a planning process, presented here in an abbreviated form that was originally prepared by Whitewood Stamps, Inc. under a grant from the Public Education Office of the United States Fire Administration (USFA), a part of the Federal Emergency Management Administration (FEMA). The complete text is available from the Public Education Office of USFA.

The five step process was developed after extensive studies of education programs that successfully reduced fire loss in different communities. Every step is based on real, successful experiences from across the nation. Although the successful programs differed greatly, they all shared an underlying systematic approach.

Public fire education is a key component of overall community fire protection master planning. The success of both public education and master planning depends on citizen awareness and community participation.

Public Fire Education Planning combines two essential ingredients: the planning **Process**, and the **People** who complete the process.

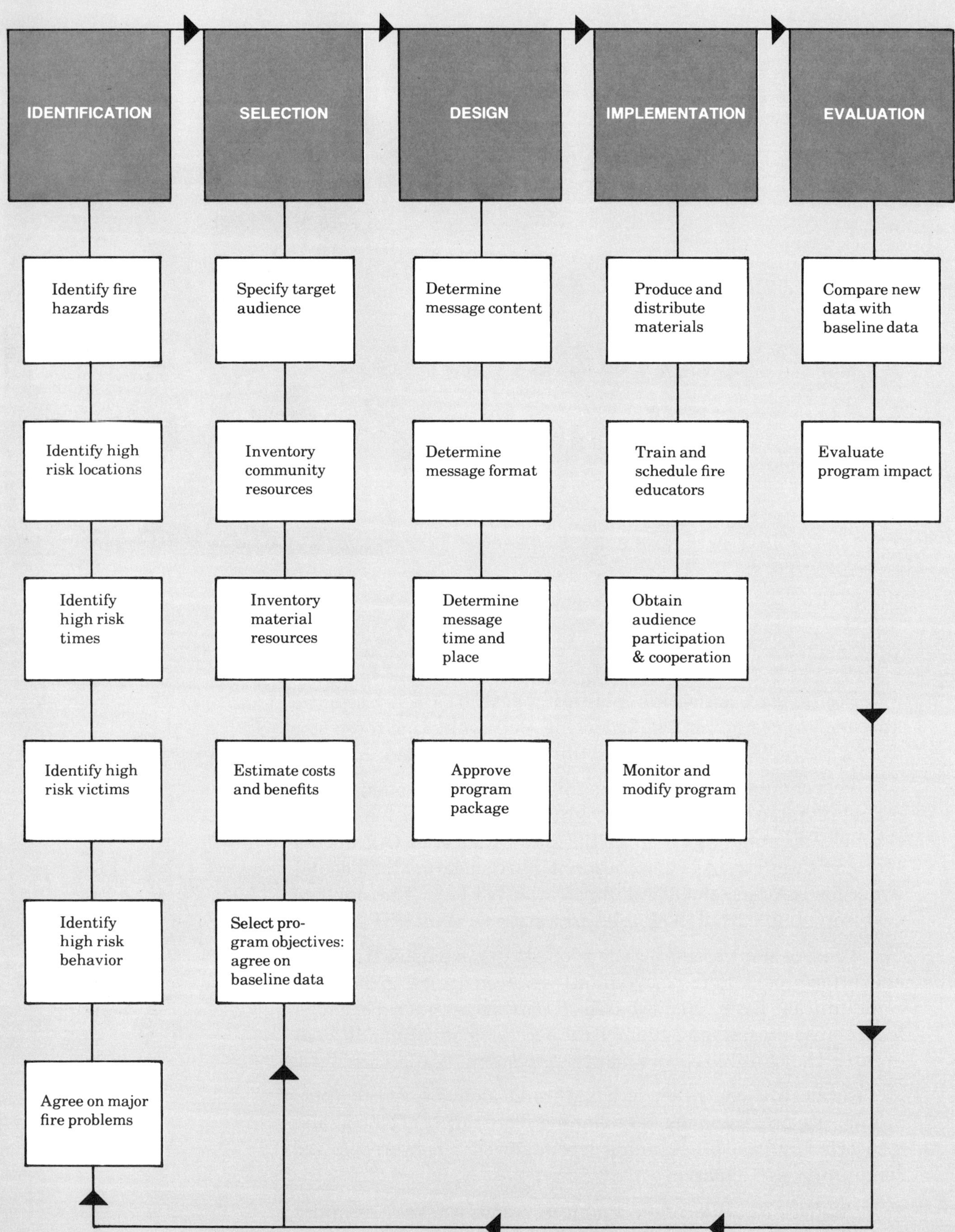

IDENTIFICATION
SELECTION
DESIGN
IMPLEMENTATION
EVALUATION
Identify fire hazards
Specify target audience
Determine message content
Produce and distribute materials
Compare new data with baseline data
Identify high risk locations
Inventory community resources
Determine message format
Train and schedule fire educators
Evaluate program impact
Identify high risk times
Inventory material resources
Determine message time and place
Obtain audience participation & cooperation
Identify high risk victims
Estimate costs and benefits
Approve program package
Monitor and modify program
Identify high risk behavior
Select program objectives: agree on baseline data
Agree on major fire problems

THE PROCESS

The planning process has a starting point and five basic steps: Identification, Selection, Design, Implementation and Evaluation. Within each step are *Activities* and a *Decision*. The *Activities* involve gathering and organizing information, materials and resources. In the *Decision* phase, the information is reviewed, summarized and acted upon. Frequently the information, when displayed, will present a clear direction which the whole group can readily agree on because it represents the consensus of all their thoughts. In the end, clear objectives should be stated.

Starting Point: Establish Responsibility

A clear understanding of responsibilities is crucial to effective planning. There are two major areas of responsibility to be established: administrative or policy responsibility and staff or program responsibility. After deciding which people are responsible for which duties, the planning process can begin.

Step One: Identification

The Identification step involves identifying the most important local fire problems so that the fire education effort can focus on those specific problems. The *Activities* include gathering information on subjects ranging from the most frequent location of fires to high risk behavior. The *Decision* is to agree on the major fire problems in your community. Once the major local fire problems have been identified, effective solutions can be developed.

Step Two: Selection

While Identification defines a community's needs for fire education, Selection is an inventory of community resources available to meet those needs and selection of achievable objectives.

The *Activities* of Selection include conducting an inventory of community resources, available materials and potential audiences, as well as estimating costs and benefits of different educational strategies. The *Decision* is selecting program objectives that meet your community's needs and resources.

Step Three: Design

The Design step moves the fire education process from planning toward implementation. The *Activities* of Design involve determining the specific content and format of fire safety messages and packaging the program for delivery to the community. The *Decision* is to outline and approve the education program package.

Step Four: Implementation

The fourth step is implementing the education program designed for your community's specific needs and resources. Implementation *Activities* include producing and distributing materials, training personnel and involving target audiences in the education process. In addition, an organization or individual will monitor the program for smooth day-to-day operation. The *Decision* is an agreement as to exactly how the fire education program will be implemented, monitored and, if necessary, modified in your community.

Step Five: Evaluation

The final step is measuring the impact of the fire education program. Among the *Activities* are comparing baseline and new data on fire deaths, injuries, property losses and incidents. You will also compare old and new information on awareness, knowledge and behavior in the community. The *Decision* is to review the program's impact and determine how to show the success of the program or how the focus of the program might be modified for better results. In this way, Evaluation returns you to the Identification or Selection steps to adjust your program.

Worksheets have been included to assist you in completing the planning process (See pages 32 - 34).

THE PEOPLE

The way people complete the process is also important. Each step of the planning process has been carefully designed to help people use the maximum resources available. During each *Activities* phase, you should list on the worksheets as many different ideas and opinions as possible. Judgments should be deferred until all the information is in. Different viewpoints should be encouraged; they may lead you to fresh approaches which will be the key to a successful program.

When working as a group make sure that each person expresses his or her viewpoint in turn. Don't be satisfied with just one point of view from the group during this phase.

STARTING POINT: Establish Responsibility

- Assign administrative and policy responsibilities.

- Assign staff responsibilities.

- Determine specific responsibilities for *Decisions* and *Activities*.

There are two kinds of responsibility in fire education planning. The first is administrative and/or policy responsibility which can include: Providing a statewide perspective, coordinating efforts with other agencies or organizations and gaining the support of influential agencies, organizations or individuals.

The second responsibility involves "front-line" or staffing tasks. Included are: Gathering information, locating or preparing materials and carrying out the education program.

You may choose to have two separate groups be responsible for the two kinds of tasks. One group, the Fire Education Committee, may have administrative/policy responsibilities while another group, the Fire Education Planning Team, has staff responsibility. Another option for utilizing both groups is to assign *Activities* to the Fire Education Planning Team and *Decision* to the Fire Education Committee.

Some fire educators may find themselves fulfilling both the staff and administrative/policy roles. Whether these two functions are performed by two groups or by one, it is important that the staff duties of the Planning Team and the administrative/policy duties of the Committee be carried out.

Broad representation from the community is a cornerstone for the success of both groups. Fire education must be a community effort, which includes fire service, schools, community service groups, and the local media to help develop community support for fire education. In particular, people from key community groups and the target audiences have influence and information to contribute to the education process.

Choosing the right people to plan and implement a public fire education program will, to a great extent, determine the success of the program. Choose the people and assign responsibilities before actually beginning the five step process.

STEP ONE: IDENTIFICATION
Objective: To Identify Major Local Fire Problems

Every successful public fire education program begins with identifying the most important local fire problems. Once this is done, all the necessary resources of the community can be used to solve those specific problems. There are many ways to obtain the information necessary to identify major fire problems. The USFA's National Fire Incident Reporting System (NFIRS) is, for example, a useful resource. Purely local efforts can also be successful.

In Seattle, Washington, for example, a census-tract-by-census-tract analysis of fire incidents revealed that the city's major fire problem was greatest in a few specific neighborhoods. As a result, home inspections and other educational programs were concentrated in those neighborhoods.

This approach was not only effective, but more economical than scattering fire department resources throughout the city. This is the type of approach recommended throughout this chapter. Identification of specific fire problems is the first step toward an effective solution.

RECORDS INDICATE HAZARDS

Activity: Identify Fire Hazards
- Locate records showing causes of fires.
- Select most frequent causes of fires.
- Determine local patterns of fires.

Records from NFIRS, the State Fire Marshal, local fire departments and hospitals will help the fire educator determine sources of ignition, such as smoking materials or heating appliances, involved in fire incidents.

This information is important because it can identify the specific local patterns of fires which should be addressed by your public education program. In Mississippi County, Arkansas, for example, the Identification phase of a fire education project revealed bad wiring and overloaded circuits as hazards in that county. Those responsible for the education program focused community attention and effort on electrical problems, reducing fire losses by 50 percent. When the most important local fire hazards are identified, a program can be directed toward these specific problems with a higher probability of achieving measurable loss reduction.

Activity: Identify High Risk Locations

- Locate neighborhoods or building occupancy types with high fire rates.

- Discover what is causing rates to be above average.

- Concentrate programs and personnel in high risk locations.

Certain neighborhoods can be "high risk" in terms of fire dangers. This is especially true in urban areas where deterioration or "blight" has set in. These neighborhoods can be identified by plotting the locations of fire incidents on a map, as in New Orleans, where fire educators found "migrations" of fire incidents within the city.

In Chicago, the "Operation Pride" home inspection program operates in the three census tracts with the city's highest fire incident rates. Where rubbish was a major Chicago fire hazard, the fire department cooperated with the department of sanitation to have special pickups in the high risk areas. This approach requires that fire educators learn a great deal about local community needs and problems.

In identifying high risk locations, it is helpful to remember that:

1. "High risk locations" can be neighborhoods or building occupancy types with high fire rates (the number of fires per capita) or high fire incidence (the absolute number of fires).

2. Fire educators working in high risk locations confront other community problems, in addition to a high fire risk.

3. Flexibility and cooperation with other city agencies will help fire educators achieve their mission.

Determine which neighborhoods or building types have the highest rate of fires and identify these locations on a map.

Activity: Identify High Risk Times

- Identify times of day or year with highest fire incidence and losses.

- Identify types of fire occurring at these times.

- Plan to concentrate fire safety messages at these times.

Common sense tells us that a Christmas tree fire safety campaign should be launched in December and not in July. In the same way, a radio spot about cooking fires should be broadcast in the early morning or early evening, when most of these fires occur.

Fire dangers increase at certain times of the day and of the year. In some communities, the most serious fire casualties may occur *only* during high risk times. Local fire educators should identify these high risk times, then organize programs that will alert people to times of high fire danger.

Los Angeles County is a good example of using knowledge about high risk times. During dry spells, the fire department dispatches apparatus to strategic places when school is dismissed in the afternoon. This action cuts down on brush fire incidents, which are often started by children on their way home from school.

"High risk times" can become moments of opportunity if the right actions are taken then.

Activity: Identify High Risk Victims

- Identify groups with high fire death and injury rates.

- Determine why they have a high fire rate.

- Involve these groups in the fire education effort.

Fire is more hazardous to some people than to others. The very young and the very old, for example, suffer more fire fatalities proportionately than any other age group. Transient residents are frequently more careless and fire prone than homeowners.

The identity of high risk victims depends on the local situation. When the people at the greatest risk have been identified, programs directed to them can be designed. The potential victims can be involved in the fire education effort.

A demonstration program in Robeson County, North Carolina, for example, found that native Americans were the most frequent victims of flammable liquid fires. A program was designed to meet their needs, and the burn injury rate dropped by 65 percent.

High risk victims — and those people who are close to potential victims — greatly need fire education, but in any given community, they must be identified before education and communication can begin.

Activity: Identify High Risk Behavior
- Determine which behavior — acts or omissions — causes fires.
- Decide how the behavior can be changed.
- Teach the people exactly what to do.

A person's actions or omissions, either before or after ignition, are often the most important single factor in a fire incident. Discovering these behavior patterns is often difficult, but the information can prove very valuable to the fire educator.

After obtaining this information, the fire educator will decide where to intervene in the dangerous behavior pattern. In the South, for example, children standing too close to space heaters and igniting nightgowns has long been a problem. Here, intervention can occur by educating parents and children about the danger, considering the purchase of flame resistant clothing, and/or teaching children what to do if their clothing ignites.

Fire educators realize that fires are not caused by flammable liquids or smoking materials. Fires are caused by the way people deal with these materials and ignition sources. For this reason, fire education programs cannot afford to overlook the relationship between human behavior to fire and burn hazards.

Identify high risk behavior and decide how the behavior can be changed. If flammable liquids are a major source of fires, teach people how to properly and safely store and use them.

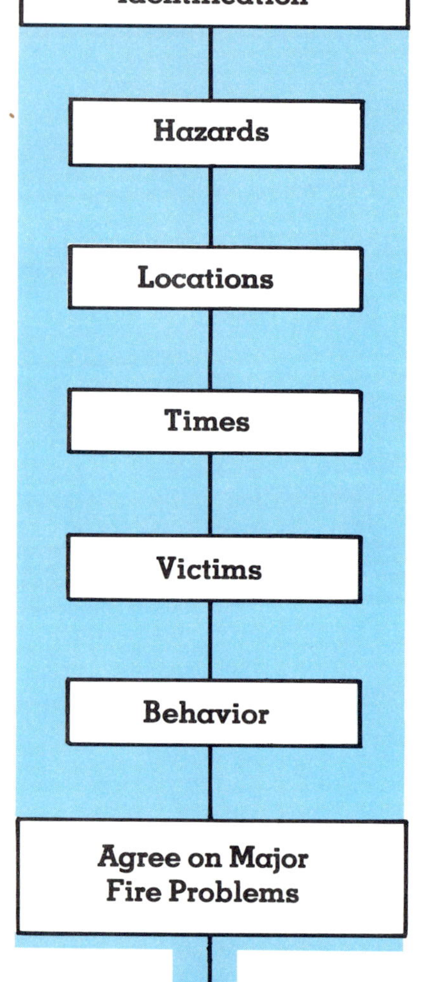

Decision: Agree On Major Fire Problems

- Review information on major fire problems.
- Determine which local fire problem is the most serious.
- Create "scenarios" of how fire happens.

The information gathered during Identification will probably reveal several serious fire problems. The task now is to choose the problem which, if solved, would have the greatest impact on the local fire situation. The fire educator may wish to determine which problem is the most serious in terms of (1) fire incidence, (2) fire deaths, (3) fire injuries and (4) fire loss.

The next step is to create a "scenario" or mental image of how the fire happens. The scenario can be made from the information gathered during Identification activities, from a detailed study of local fire data, or from an in-depth investigation (including human behavior) of selected fire incidents. The scenario will help identify target audiences and strategies for reaching them.

Hypothetical scenarios include:

1. "Children aged 10 and below (high risk victims) setting fires (high risk behavior) in apartment trash rooms, especially in the northeast section of town (high risk location), after school hours (high risk time)."

2. "Adult females (high risk victims) leaving the kitchen unattended during meal preparation (high risk behavior and high risk location), especially when preparing breakfast in the morning (high risk time)."

3. "Adult males (high risk victims) smoking in bed while under the influence of alcohol (high risk behavior and location)."

STEP TWO: SELECTION

Objective: To select the most cost-effective objectives for your fire education program.

When the most serious local fire problems have been clearly identified, a strategy to solve those problems can be selected. Depending on local resources, the fire education strategy might emphasize mass media, in-school programs or community-wide fire education. The choice will ultimately depend on the target audience you have chosen.

In Cincinnati, for example, the Shriners' Burns Institute recently concentrated on training teachers in fire safety. This strategy was aimed at having maximum impact for minimal cost by reaching many children through a few teachers. In Los Angeles, where it is not feasible to reach all the schools, the fire department emphasizes working through the local television stations.

The basic idea is to review the local situation and to choose the strategy with the greatest potential effect at the lowest cost.

In selecting the target audience, remember that children will very likely take home some of the ideas presented to them. In this manner, children and their parents may become more fire safety conscious.

Activity: Select Target Audiences

- Review the high risk victims identified in Step One.
- Identify those who influence the high risk victims.
- Select the audience with the greatest potential impact.

Selecting the target audience for your fire education program can be the key to its success. It is important to remember, however, that the target audience may not be the people identified as high risk victims in Step One. Those who influence the potential victims are a possible audience. For example, school children, the school board, the superintendent, the principal or a teacher may also be your audience.

In working with the elderly, you may decide that a program aimed directly at retired persons or at their social organization is the best approach. On the other hand, you may decide that the elderly related fire problems are tied to their housing environment and that your target audience should be people who can pass and enforce codes or managers and nurses at local health care facilities.

Many local stores will allow fire prevention literature to be distributed or posters to be displayed in their buildings. Some owners/managers may also provide materials for constructing such literature at a reduced or no charge basis.

Activity: Inventory Community Resources

- Identify influential people in the community.
- List all local media and civic organizations.
- Make personal contact with key people and groups.

A "community resource" is anything or anyone that can transmit or help transmit fire safety messages to the public — local television and radio stations, newspapers, civic groups, schools and community organizations, and influential individuals. Once these community resources have been identified, the most effective ones for reaching your target audience can be chosen for the fire education program.

In Santa Ana, California, for example, the local Kiwanis Club was one of the first organizations to support public fire education. In Boston, a local radio station has recently run a series on arson. In Erie, Pennsylvania, the Jaycees are conducting a smoke detector program.

In developing an inventory of community resources, the fire educator should remember that:

1. Local broadcasters are required by the Federal Communications Commission to render some services to the communities in their license area. Station managers will often help a well-planned fire education project.

2. Also, local civic groups seek worthwhile community projects and have traditionally been sources of funds and people for fire education projects.

3. Effective communications is a two step process. Messages transmitted through the media have the greatest effect when reinforced by local community leaders.

Local civic organizations are usually receptive to any project that will benefit the community. Soliciting their support can be extremely beneficial to a public education program.

Activity: Inventory Material Resources

- Ask businesses and organizations what materials, equipment or skills they could donate to your programs.

- Determine availability and cost of fire education materials.

- Review existing programs and determine the advantages of purchasing educational materials versus making your own.

Most communities have the resources to create the fire safety materials needed to conduct a public fire education program. An inventory of these resources should determine both availability and cost to the planned fire education program. In Guilford County, North Carolina, for example, a local business supplies decals to the County Fire Marshal's office for about one cent as opposed to the 5-10 cent commercial rate.

In conducting the inventory of material resources, the local fire educator should remember that:

1. Large companies often employ highly skilled graphic artists and other professionals whose time is frequently loaned to community projects as a public service.

2. Community shopping centers, fast food stores and civic clubs are always looking for interesting projects to sponsor.

3. Schools and other organizations often have audio-visual equipment (such as slide projectors) which can be used for fire education projects.

Descriptions of coummunity-developed education programs are available from the Public Education Office, USFA. Excellent commercial materials are available from several sources, including the National Fire Protection Association, Film Communicators, Inc., and the Hartford Insurance Company.

Large companies may allow their artists to design fire prevention literature or offer their services in other areas. It is the responsibility of the fire prevention educator to solicit such assistance.

Activity: Estimate Costs and Benefits

- List alternative program objectives.

- Estimate costs of alternative program objectives.

- Estimate loss reduction impact of each program objective.

- Choose the most effective approach within limits of local resources.

The most effective fire education programs are targeted campaigns based on clear objectives and program strategies. Comparing expected costs to expected benefits will help in choosing the best strategy.

A cost-benefit analysis can demonstrate the effectiveness of public fire education. In Louisiana, for example, it was estimated that the Forest Service's "personal contactor" program prevented 310 fires a year at a cost of $24 per fire — much more than the cost of the program.

Cost-benefit estimates will not tell the fire educator which objective or strategy to choose, but will demonstrate the probable impact of different choices. For example, residential fires usually cost less than industrial fires in terms of property losses, but more lives are lost in residential fires. If it costs the same to prevent both types of fires, which would you, as a fire educator, emphasize?

Decision: Select Program Objectives

- Agree on specific and attainable educational objectives that are compatible with local resources and people.

- Agree on baseline data.

- Present a clear "image" of program objectives around which to mobilize the community.

Since a specific local fire problem was identified in Step One, selection of program objectives should focus on specific solutions to this problem. The objectives should be attainable within the limits of time, money and available personnel. Wherever possible, the objectives should be measurable.

The analysis of costs and benefits should point to the best objective. If you find that you cannot solve the most serious problem with the resources available to you (i.e., you cannot attain your objective), you may revise your objective or return to Step One to identify the next most serious fire problem.

Agreeing on the baseline data — the extent of a specific problem at the beginning of the education program — is an important part of Selection. Both loss data (the statistical extent of a problem) and educational data (surveys of appropriate fire safety knowledge) can be part of your baseline data on a par-

Make certain the benefits of meeting the program's objectives outweigh the costs of implementation. Benefits do not necessarily have to be measured monetarily.

ticular problem. This baseline figure will be carefully monitored during the program. A change from the baseline, measured during Evaluation, is one of the best indications that a public education program is, in fact, working.

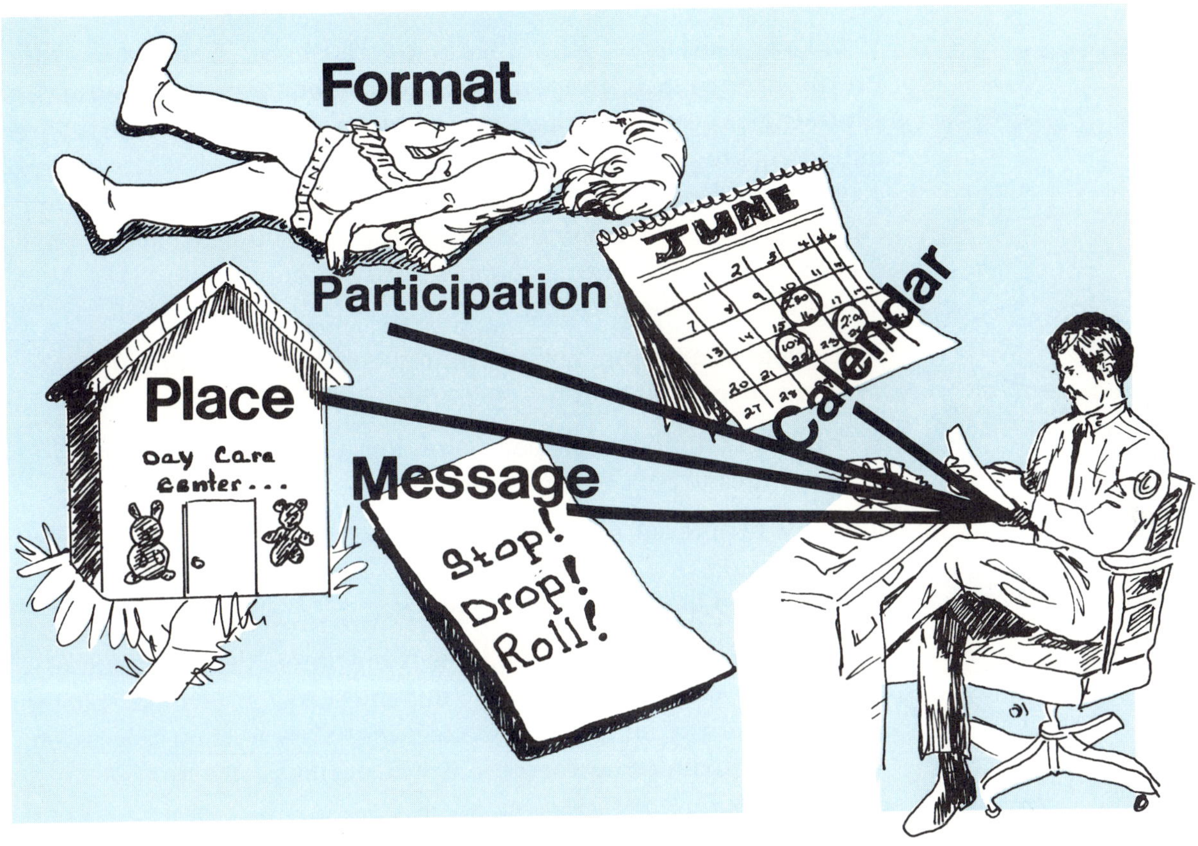

In designing a program, many decisions must be made: What is the message to be conveyed? Where and when will the message be distributed? How can the message best be presented? All of these questions must be specifically answered and planned in detail for the program to be successful.

STEP THREE: DESIGN

Objective: To design and develop effective education program materials.

The Design step is the bridge between planning a fire education program and implementing it in the field. This is the stage when final decisions are made about specific formats, messages and times for distribution to the target audiences. At this time, the fire educator decides what to say and how to say it, and puts the program into a package that can be used by everyone involved in the effort.

Public fire education programs across the country have successfully used local talent to design and produce materials. From the Action for the Prevention of Burn Injuries to Children (APBIC) Burn Prevention Program in Boston to the Preschool Program in Stillwater, Oklahoma, to the "House of Hazards" of Prince Georges County, Maryland, local fire educators have shown great creativity in designing effective materials.

Activity: Determine Message Content

- Direct messages toward specific hazards.

- Appeal to positive motives.

- Show the context of the problem and desired behavior.

The content of the messages must be directed to the elimination of specific fire hazards identified in the first step of the planning process. The key principles of designing effective fire safety messages were investigated in "A Study of Motivational Psychology Related to Fire Prevention Behavior in Children and Adults," for the National Fire Protection Association's "Learn Not to Burn" campaign as popularized on television by Dick Van Dyke. These principles are:

1. Appeal to existing motives that people have for being fire safe; don't threaten them.

2. Be explicit about proper behavior and specific in directing messages to particular groups.

3. Show people the situation you are talking about and tell them what to do.

The NFPA's "Learn Not to Burn" program has been based on these principles, and it has had as remarkable degree of success. Numerous cases have been documented in which people remembered the simple, clear messages they saw on television, and then responded correctly in fire situations.

There is increasing evidence that public fire education does work when people are taught how to recognize fire hazards, how to correct them, and how to react in fire situations.

Keep message content positive.

Activity: Determine Message Time and Place

- Determine when the target audiences will be most receptive to fire safety messages.

- Schedule messages for maximum effect.

The inventory is intended to help the fire educator determine the best times and places to schedule messages for target audiences. For example, smoke detector advertisements in Los Angeles are scheduled for prime time evening hours when adult viewers (potential purchasers) are most likely to be watching television. In Miami, home inspections are scheduled for Saturday mornings, when people are at home, but before they begin watching sports events on television.

In conducting this inventory, the fire educator should consider the following:

Activity: Determine Message Format

- Match format to message.

- Match format to audience.

- Match format to resources.

The best format for the selected messages depends on what is being communicated, to whom it is being communicated, and the resources available to the local fire educator.

Showing the correct behavior in a fire situation, for example, might best be done as a demonstration or on television. A printed home hazard check list is helpful for locating fire hazards in the home. A simple wall poster with strong graphics is best in commercial or industrial establishments. Filmstrips are valuable for teaching preschoolers with short attention spans. Longer films are effective in high schools.

The Massachusetts Firefighting Academy decided that teaching firefighters basic smoke detector information was the best way to reach the public, since firefighters often handle telephone inquiries about smoke detectors. The most appropriate format for their program was a self-instructing technical manual which could be used by students in a course and as a reference book in the station.

Ultimately, available resources are the determining factor in choosing a format which is both affordable and effective.

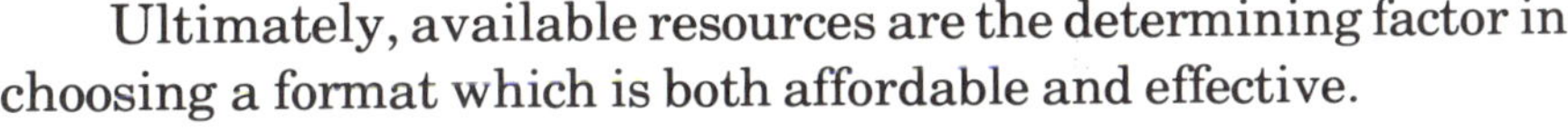

1. Most television and radio stations subscribe to rating services which tell them when different types of people are watching or listening to specific programs. Frequently, the ratings' data will be made available for planning fire education.

2. People are receptive to different types of fire education messages at different times in their lives and at different locations during the day. Take advantage of these facts by placing pamphlets on fire safety for children in the waiting rooms of pediatricians' offices or dropping leaflets about cooking fires in bags at supermarkets.

3. The best sources for information on how to reach specific groups are representatives for the groups themselves. Ask them when and how to reach their friends and neighbors.

Decision: Approve Program Package

- Design program package.

- Determine how materials will be produced.

- Present planned materials to sample audience.

After making final decisions about the format, content and timing of the fire education messages, you must design or purchase the educational materials. This means putting the educational message into the form of slides, videotapes, posters or brochures which will be most effective in your community.

In cities or towns with easy, direct contact between fire educators and their audiences, simple portable media like slide shows are often preferable. In larger cities, where direct contact is limited, packaging for the mass media may be more effective.

Once the materials have been designed or purchased, it is important to present the material to some people in the target audiences. These people will critique the material and you can observe their reactions.

The most effective program materials package must be adapted to local audiences and local problems. When you are confident that the materials have been tested and will work, the implementation phase can begin.

STEP FOUR: IMPLEMENTATION

Objective: To effectively implement the public fire education program plan.

Participation is the key to successful implementation of a public fire education program. The broad-based membership of the Fire Education Planning Team and the Fire Education Committee began local involvement in your community.

But participation and cooperation are needed from other sections of the community as well. The uniformed firefighter needs to become involved, as do the nurses and doctors in burn treatment centers, and the target audiences for prevention messages.

Fire prevention is not only a local problem, it is also a local process. Like every other activity at the local level, people's involvement is the key to the program's success.

Activity: Produce and Distribute Materials

- Assign production responsibilities.
- Produce or purchase materials.
- Distribute materials to target audiences.

The basic goal of the entire fire education program is providing fire information to the public at large, and especially to high risk target audiences.

In Los Angeles, for example, the fire department worked with the local television station to produce a four-part documentary on smoke detectors and fire in the home. In this case, production and distribution were handled by the station, with assistance from fire department advisors. In Ohio, a statewide committee on fire safety designed and produced its own slide presentation on smoke detectors and home escape procedures. The same group is now working on fire education materials for children.

In other communities, community college art classes or media classes are asked to produce educational materials. Adopting model programs tested in other communities is another very effective solution. Frequently, civic groups or fast food chains will produce and distribute fire education materials as a service to the community.

Many communities also purchase their educational materials from the National Fire Protection Association (NFPA) or participate in the Hartford Junior Fire Marshal Program.

Plan for individuals to both produce and distribute the material that is to be presented to the public.

Activity: Train and Schedule Fire Educators

- Organize fire service personnel and volunteers.

- Train people for their job.

- Match community "contacts" with target audiences.

Where do you find the people to implement the fire education program? There are many possible answers to this question, each one growing out of specific local circumstances.

In Mt. Prospect, Illinois, public education is considered a part of the uniformed firefighter's job. Fire service personnel perform a number of public education activities throughout the year, with a very positive effect on the city's fire rate.

In Edmonds, Washington, civilians were hired with Comprehensive Employment Training Act (CETA) funds to conduct a Home Safety Survey Program, a home inspection effort which has reduced property losses in Edmonds by 69 percent. When CETA funding became unavailable, the fire department recruited and trained elderly volunteers to survey Edmonds' homes. In Delaware, the State Fire Training School has organized teachers, preschool specialists, nursing home personnel and wives of volunteer firefighters to carry out community fire education. The school has designed and taught two-day courses for fire safety instructors.

Examples such as these can be found in many communities which are implementing successful public fire education efforts. People can be found, motivated and trained to teach fire safety courses which they have never taught before — and to do it very well.

A fire educator does not necessarily have to be someone directly affiliated with the fire service. Survey the community and solicit the assistance of interested civilians to help spread the message.

Activity: Obtain Audience Participation and Cooperation

- Involve target groups in implementing programs.

- Tell target audiences what to expect.

- Reinforce messages through endorsement by local opinion leaders.

The selected audiences for public fire education programs should always be active participants in the education process. Early participation by these groups helps insure acceptance of the education program, rather than creating resistance to it. In Tampa, Florida, for example, the fire department organized senior citizens into mutual assistance for fire safety brigades.

In Edmonds, Washington, homeowners received letters with their water bills notifying them that the fire survey team would be visiting them soon. The notice also requested the recipients' cooperation in the survey. The result was a near-90 percent entry rate into Edmonds' homes, a very high percentage for home inspection programs. Explanations and endorsements by a Spanish-speaking priest increased the entry rate in a similar program in Chicago's Spanish-speaking neighborhoods.

Fire educators should remember that effective communication is often a two step process. Messages received through the mass media are far more effective when reinforced by opinion leaders respected in the high risk group or by the community as a whole. Participation and cooperation by the community at every possible level is a fundamental principle of Implementation.

If home surveys are to be conducted, use the mail to alert the community as to the purposes and need for the upcoming surveys. Gaining support before actually making the surveys will make the job much easier.

Decision: Monitor and Modify Program

- Observe program operation on a day-to-day basis.

- Modify program on the basis of the review.

Day-to-day monitoring of the program during Implementation permits routine adjustments to be made. A more formal review should be undertaken at intervals of six months or more so that more substantial changes can be made.

In the case of "Operation Pride," the Chicago Fire Prevention Bureau, in cooperation with the local sanitation department, began a neighborhood clean-up program. The plan was for voluntary home inspections to be followed a few days later by extra garbage pickups. When the sanitation department was unable to provide trucks to meet the enormous public response, the Bureau modified the program by concentrating on the fewer neighborhoods with the highest fire incidence rates.

The ability to be flexible and to change a program after it has begun is an essential characteristic of the public fire educator.

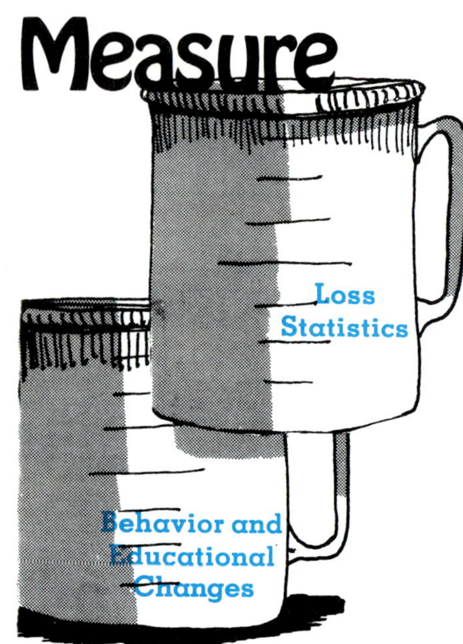

STEP FIVE: EVALUATION

Objective: To measure the impact of the public fire education program and modify the program if necessary.

Evaluation provides the planners with information once the program is underway. If the program has been effective, more resources can be committed. If not, the program can be revised.

In Louisiana, for example, sociologists working with the U.S. Forest Service found that programs using the mass media effectively communicated information, but did not influence behavior. They found that their target audiences required face-to-face reinforcement of the fire message before they would respond to it. The program was re-designed, using a "contactor" approach. Evaluation showed a 55 percent reduction in set fires in one year.

Evaluation is a very valuable tool. It not only measures impact, but also feeds back information so that program modifications can be made to insure future success.

Activity: Compare New Data With Baseline Data

- Make new measurements.

- Compare with baseline loss data.

- Compare results with overall program objectives.

What is the impact of public fire education? Comparing loss and educational data before and after implementing education programs provides a generally reliable measurement of the effectiveness or impact of education.

One method of measuring program impact is comparing loss statistics (death, injury, incidence and property loss) before, during and after an education program. For example, death declined between 43 percent and 63 percent in pilot projects conducted by the U.S. Department of Health, Education, and Welfare in Missouri, Arkansas and North Carolina. In Santa Ana, California, a 60 percent reduction in per capita fire loss has been recorded. The USFA's NFIRS System is a valuable tool in making this type of comparison.

Evaluating a public fire education program should also include measurement of changes in fire safety awareness, knowledge and behavior. Ways of measuring these "educational changes" include:

1. Comparing scores of questionnaires before and after a program. This not only provides information to the fire educator, but also reinforces the public's increased knowledge.

2. Conducting telephone surveys of those attending presentations to determine how behavior has changed, e.g., purchasing smoke detectors.

Changes in fire risk, such as the number of hazards per home, can also be measured.

This new data will not only illustrate the impact of the education program, but can also serve as baselines for future programs.

Decision: Evaluate Program Impact

- Summarize results from all evaluation.

- Decide how to change and improve program.

- Return to Identification or Selection phase.

A careful evaluation will point to very specific ways of improving a public fire education program. These suggestions for improvement can then serve as clear recommendations for further planning and implementation.

The following are some hypothetical examples:

1. "Since our campaign began, deaths from careless cigarette smoking have been cut in half, which was our objective. Recommended that we now begin to concentrate on space heaters."

2. "Knowledge of flammable liquid dangers has increased since the program was implemented, but behavior has not changed. Recommend we try a 'personal contactor' program."

3. "Objectives for distribution of material are not being reached. Recommend we hire more personnel."

Once the Evaluation phase has produced new, specific recommendations, the fire educator returns to Identification for a new problem or to the Selection phase for new education strategies, correcting areas of weakness and supplementing areas of strength.

The entire process has now been completed and begins once again.

WORKSHEET 1 - IDENTIFICATION

Objective: To identify major local fire problems.

Activities: Answer the following questions about your community's fire problems.

What are the major fire hazards?

- *Careless smoking*
- *Kitchen grease fires*

Where are the high risk locations?

- *Northeast section of town*
- *Bat. No. 3*
- *Sudbury Apts.*

When are the high risk times?

- *Weekends*
- *Nights (10 p.m. - 2 a.m.)*
- *Kitchen — early evening*

Who are the high risk victims?

- *Adult males (smoking)*
- *Adult females*

What is the high risk behavior?

- *Smoking and drinking*
- *Leaving the stove*

Decision: Agree on major fire problems in your community.

Residential fires started by smoking/drinking, especially in Bat. No. 3 area; usually at night and on weekends.

WORKSHEET 2 - SELECTION

Objective: To select the most cost-effective objectives for your education program.

Activities: Answer the following questions about your community.

Who are your potential audiences?

- *People who live in Battalion No. 3 area*
- *Adults, especially males*
- *Apt. residents*

What are your community resources (civic groups, media etc.)?

- *Lions, Kiwanis, K of C*
- *Sudbury Tenants Association*
- *Cement Plant (employs Sudbury residents)*
- *WPFE Radio*

What materials can you use?

- *Smoke detector film*
- *Smoke detector brochure (from USFA)*

Estimate costs and benefits of the various education programs.

- *Reprint 5,000 copies of brochure for $500*
- *New slide show - $500*
- *Now 12 runs a month to Sudbury Apts., 78 smoking fires*

Decision: Select achievable objectives.

- *To reduce smoking fires in Sudbury by 50% in 6 months*
- *To have residents install 500 smoke detectors in six months*
- *Use 78 smoking fires as baseline*

WORKSHEET 3 - DESIGN

Objective: To design and develop effective educational program materials.

Activities: Answer the following questions about your community.

What is your primary message?

- *Smoke safely — buy a smoke detector*

Which formats are best to present your message?

- *Presentations to local groups*
- *Distribute brochure*
- *Show slide show or movie (depending on the group)*

When are the best times and places to present your message?

- *Just before payday — so people can buy detectors soon*
- *Scheduled meetings of local groups in the neighborhood*

Decision: Specify your education program package.

- *Smoke detector presentations at meetings of local groups*
- *Reinforce message with slide show or movie*
- *Send brochures home with the audience*

WORKSHEET 4 - IMPLEMENTATION

Objective: To effectively implement the public fire education plan.

Activities: Answer the following questions about your education program.

How will you produce your materials?

- *Sgt. Miller to make new slide show*
- *Lt. Thomas to get brochure printed in city shop*

How will the materials be distributed?

- *As part of presentations*
- *Scout Troop 48 to distribute additional brochures.*

Who will train and schedule the fire educators in your community?

- *Sgt. Miller is responsible*

How will you obtain participation and cooperation from target groups in implementing the education program?

- *Lions' President to form committee*
- *Mayor Jones to endorse detectors on Radio WPFE "Spot" PSA*

Decision: State how the public education program will be implemented and monitored in your community.

- *The program will be implemented in 6 weeks, with the help from Lions Club, Troop 48, etc.*
- *Sgt. Miller to keep chart of all contact with the public on smoke detectors (when, where, who, any problems)*

WORKSHEET 5 - EVALUATION

Objective: To measure the impact of the public education program and modify the program if necessary.

Activities: Answer the following questions about your community.

How will you measure reduced fire deaths, injuries, loss and incidence?

- *Bat. No. 3 loss will be compared with baseline after 3, 6, 9 and 12 months*

How will you measure changes in knowledge, awareness and behavior?

- *Community poll to see how many smoke detectors are installed*

Decision: State the impact of the public education program in your community.

- *Fire incidence down to 32*
- *800 detectors installed*

Determine to return to Identification (to define new problems) or to Selection (to select new strategies and objectives).

2
Target
Age
Groups

The most dangerous fire and burn hazard in the country is the attitude that "It won't happen to me." We know, however, that it can and does happen at the rate of thousands of deaths and burn injuries yearly. Because this attitude does prevail, one of the primary responsibilities of a fire department is to acquaint the public with the fire problem and then to educate them how to avoid the tragedies of fire.

A general awareness of fire should be instilled in the citizens of a community — adults and children alike. By teaching children the ways of prevention, adults of the future will be safer; by educating adults, the community will be safer today. Therefore, the fire department has a responsibility to teach people the skills which can save their lives. This responsibility cannot be fulfilled during one week in October which is designated as "Fire Prevention Week." It must be a year-round effort.

Firefighters must have an awareness of the fire problem so they can pass it on to the public. Each department member should learn everything possible to help the citizens of the community and teach them everything possible so that they can learn how to protect themselves. It is far better to meet a smiling, knowledgeable public in a fire education classroom than to meet a burned, agonized public on the fireground.

The public can be reached through various methods: the news media, schools, civic groups and other groups. Information for some of these will be given in this chapter in an effort to help the firefighter plan and present programs which will be beneficial to the public.

FIRE EDUCATION LEARNING OBJECTIVES FOR PRESCHOOL CHILDREN

A child needs to know many things about fire prevention and about the proper response to a fire or burn situation, including:

- Stop, drop and roll in a clothing fire emergency
- The proper behavior concerning matches
- Home exit procedures and skills
- Scald burn hazards in the environment
- Contact burn hazards in the environment
- Electrical burn hazards in the environment

Three-year old children are not too young to learn fire safety

STOP, DROP AND ROLL!

The first fire safety priority for young children is to know about the technique and idea behind the principle STOP, DROP AND ROLL if their clothing is on fire. The children should be able to demonstrate this behavior in a role-play activity.

Many young children are burned and injured by the ignition of their clothing. Clothing fires cause more severe burns than burns on an unclothed area. Injury from clothing fires can be greatly reduced if a child will immediately Stop, Drop and Roll.

This action protects the face, neck and breathing passages. Heat and flame rise, so that when a child is standing up they go straight into the face, nose and mouth. This is aggravated if the child runs and fans the flame, so the first requirement is to stop, fall to the ground and roll over and over. When the child drops and becomes horizontal, the heat and flame still go up, but away from the sensitive areas. Rolling smothers the fire to extinguish it.

How To Discuss Clothing Fires and Stop, Drop and Roll

- *How old are you? (Or, How old is everyone here today?) Is anyone three years old? Is anyone four years old?* Adapt these questions to the age of the group.

- *How many of you have on pants? (Or, Who has on a dress? Who has on jeans? etc.)* Discuss your own clothing.

- *Everything we have on will burn* (catch on fire). *How many of you realized that?*

- *How would your clothing get on fire?* Stove, match or cigarette, charcoal grill, heater in camper, fireplace, heater in bathroom.

- *If your clothing catches on fire, this is what you do — Stop! Drop! and Roll! Cover face with hands (Roll, "smush" out the fire).* Clap hands to emphasize smushing out fire and demonstrate for the children. **Note:** ACTUALLY DROP AND ROLL AND SHOW THE CHILDREN.

- *Everyone practice with me.*

Allowing children to actually practice Stop, Drop and Roll is the most effective way of impressing them with the correct behavior. Notice that one child is rolling while not covering her face. If the fire is on the hands or arms, the children should be taught to keep their arms to their sides while rolling.

How To Practice Stop, Drop and Roll

Practicing the skill Stop, Drop and Roll is the vital element in teaching a child to react efficiently and effectively in an actual clothing fire. Use one or more of these ideas:

Use a red rubber, velcro or plastic ball or a small bean bag. Toss it gently at a child. Instruct the children that when and where the object hits them is the place and time of a pretend fire. The child should immediately Stop, Drop and Roll.

Use a felt flame. Place it on a child's clothes (it will cling). This is a pretend fire on the child's clothes. The child should Stop, Drop and Roll and the flame will fall off. A good cause-effect illustration.

Use a red cloth or paper streamer or a ribbon. Tie it on a child's arm, leg or trunk (a clothed area) to simulate a flame. The child Stops, Drops and Rolls on the pretend fire.

Use a "pretend" paper flame on the end of a tongue depresser (each child can make his own as an activity). Touch the child's clothing with the paper flame. The child then can practice Stop, Drop and Roll.

Play a circle game. With everyone sitting on the floor in a circle, describe one child's clothing. When the child identifies himself as "it", instruct him to Stop, Drop and Roll. After demonstrating, that child can describe someone else's clothing in the circle and the activity continues. For very young children the instructor may want to describe the children's clothing, thus keeping the game at a faster pace.

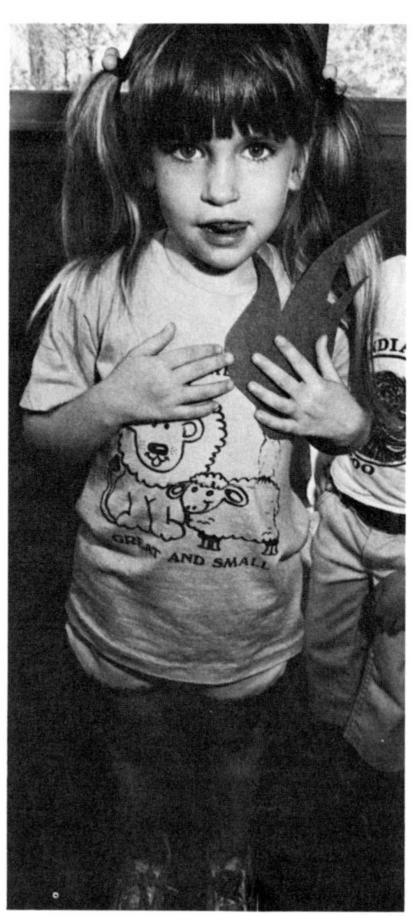

Having a felt flame to stick on the children's clothing will assist in creating a good cause-effect relationship. When teaching the children to cover their faces while rolling, make sure you place the flame in a proper location.

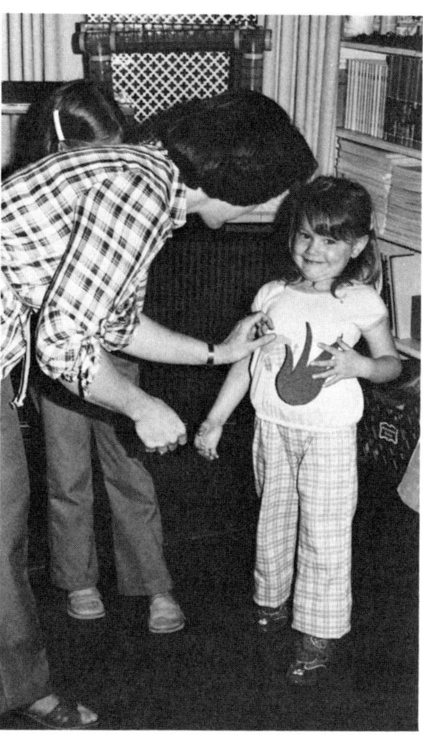

Materials To Teach Stop, Drop and Roll

Prepared educational materials are available to use when teaching Stop, Drop and Roll.

"The Story of the Little Red Fire Hat"
Fire Service Training • Oklahoma State University
Stillwater, Oklahoma 74074

Story and patterns available — Please send self-addressed and stamped envelope. This is an open-ended flannel board story. "The Little Red Fire Hat" can tell the children about Stop, Drop and Roll. Use this in conjunction with the discussion previously listed. All ages like this story (See page 136).

"Smush the Fire Out"
Film Communicators • (213) 766-3747
11136 Weddington Street • No. Hollywood, California 91601
Available in: 16mm film, 33mm slides, 35mm filmstrip

"Smush the Fire Out" is an 11-minute documentary about children having fun learning about fire. The children tell the basics of fire survival. The information includes Stop, Drop and Roll. The program is designed for kindergarten and primary grade children. Especially good for the classroom teacher who is implementing fire safety education.

"Fire and the Witch"
Film Communicators • (213) 766-3747
11136 Weddington Street • No. Hollywood, California 91601
Available in: 16mm film

Hansel and Gretel teach the Witch the fundamentals of fire prevention including Stop, Drop and Roll, extinguishing grease fires and reporting a fire. (11 minutes)

"Ban the Burn"
Action for the Prevention of Burn Injuries to Children
P.O. Box 347 • Burlington, Massachusetts 01803
100 slides and teaching manual

This is a comprehensive burn prevention program designed to be mixed and matched to meet the needs of any age audience. It contains slides of the burned clothing of children who were fire victims. Detailed instructions for using the slides are in the manual.

MATCHES

Children and matches are a universal problem. Matches, lighters and other smoking materials provide the single greatest fire hazard to young children. Children should be aware of the behavior expected of them concerning matches.

There is no single "right" philosophy concerning children and matches; instead a wide range of philosophies is promoted by various parents and educators. The most important factor is that the behavior or philosophy chosen by the parent or educator is conveyed to the children and enforced consistently. Children must know what is expected of them.

Educational Philosophies Concerning Matches and Children

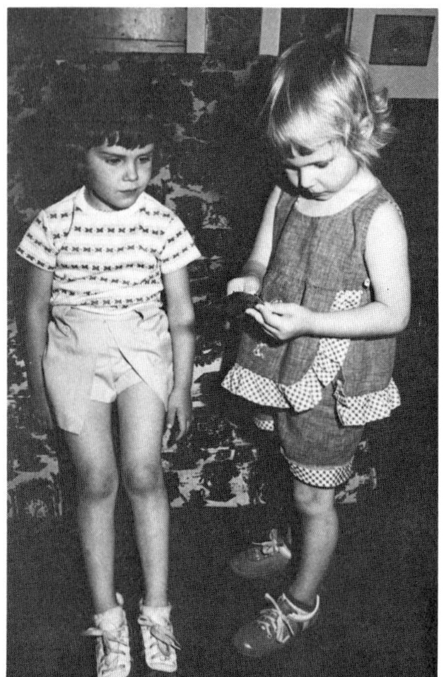

Children who are curious about matches might play with them without proper supervision. This is the reason it is extremely important to keep matches out of the reach of small children and to teach older children safe behavior concerning matches.

- *Matches aren't for children.* This approach (hands off, **never** play with matches) is most strongly supported for very young children who have no reason to use a match. These children, under six years old, usually do not have the physical dexterity to make a match work. The problem with this approach is that there is no room to explore the natural fascination with matches that children from four to six have. As a result, this approach should be accompanied with another philosophy or abandoned altogether when children show a tendency towards playing with matches. Environmental control (all matches out of children's reach and access) is an excellent accompaniment to this approach.

- *A match is a tool.* Children should be made aware of the use and function of matches and fire. The match should be presented as a tool having a specific function (lighting candles, starting camp fires, lighting a fire in a fireplace) and the use of a match for purposes other than those for which it was designed should be presented as being improper use of the tool. The match is not a toy, and to use it as a toy would be the same as attempting to drive a nail with a saw or cut a board with a hammer. A tool has a specific purpose and should be used only for that purpose. In general, the best age to begin the "match is a tool" education is at age five; however, the parents will have to determine the specific age for this activity. Prior to age five, it is considered a good idea to keep all matches, lighters and other such materials out of the reach of the child.

- *Use matches only with adult supervision.* This approach may be coupled with a "match is a tool" but takes into consideration the match fascination of young children. With this philosophy a child is free to "play with matches" but only in the presence of a supervising adult. This is well thought of combat to the problem of a child who habitually plays with matches or has even set a fire. A negative approach — do not play with matches — may only promote more curiosity and fire play while making it a secretive activity. This often forces the activity into a dangerous area such as a closet or other hiding places. Setting the boundary that matches may be used only in the presence of the parents or supervising adults and soliciting an agreement or promise from the child may be the simple solution to match-play problems. The parents' cooperation is a necessary element in this approach. They must agree to provide the supervision and allow the child to use matches when the child expresses the desire.

- *Match use saturation.* A child's curiosity is generally what causes the child to play with matches. A match offers considerable fascination to most children and there develops a desire to investigate this mysterious object. One minute you can carry it in your pocket and next you can produce fire, heat and light. In some cases this curiosity grows out of control and reaches an extreme point where the child and his environment are in danger. When this situation arises it is suggested by some parents and educators that the curiosity should be saturated under guidance circumstances.

Place the child in a safe, fire-resistant environment and remember the child's clothing offers the greatest burn hazard. Demonstrate the proper way to use a match by opening the package and taking out a match. Close the package and use the striker surface on the package. Hold the match in the proper position and strike away from the body. After ignition hold the match horizontally. Finally, blow out the flame and wait for the match to cool before disposing of it.

After showing the proper way to light and dispose of a match, allow the child to strike the matches. Repeat the sequence. It is recommended that the child should strike 200 matches or a large box of matches before being allowed to stop. Generally after 30 matches the child will not wish to continue. The child should be forced to continue until 200 matches are struck or until he absolutely refuses to strike another match.

Instruct the child to dispose of all the match debris properly. When the child reaches this "saturation point", he will usually not be interested in playing with matches any more.

This "200 match sequence" has a double effect: it saturates the child's curiosity about matches; and the match no longer is looked upon as a "fun" thing, since the parents forced the child to continue striking matches. Many activities are fun until one is forced to perform the activity, and then the activity becomes work.

It is important to take only one child at a time through the saturation routine. One child may try to outdo another with respect to number of matches struck.

Match use saturation will often become work and not fun to a child. This is one effective method of satisfying a child's curiosity towards matches.

Materials To Teach Match Behavior

"Matches Aren't For Children"
>International Fire Service Training Association
>Oklahoma State University
>Stillwater, Oklahoma 74074

A 16-slide program with narrative manual, for three- and four-year olds. A pretend story about what happens when a little girl plays with matches.

EXIT DRILLS IN THE HOME (EDITH)

Home exit procedures and skills should be taught to young children and can be practiced in a classroom situation.

Children are often responsible for themselves in a fire situation. They must know what to do and how to do it. Home exit skills are physical abilities and manipulations such as unlocking doors and raising windows, procedures which a child must implement. In this learning situation positive information is especially important. In an emergency situation there is no point for confusion between right and wrong behavior. Use only "do's" and not "don'ts."

Teach children to crawl when escaping from a smoke-filled room. The "participation panel" on the right is part of the National Fire Safety Exhibit and Show sponsored by the Burger King Corporation which tours shopping malls across the country.

What To Teach

Young children live in a variety of different home environments requiring specific exit plans. There are two basic actions for children in a fire situation which are almost always correct. The first one is to get down low and crawl on the floor. The second one is to go to the window. These two points provide a solution to several problems. The most important is that they give the child an action plan to follow. This will keep children from hiding in closets and under beds, which all too often is the course of action children follow automatically. These actions also place the child in a safer environment (on the floor) and in a location to be more easily rescued (at a window).

Other more complicated procedures can also be approached with young children, such as: exiting out a first-story window, deciding on two ways out of a room, planning a meeting place for the family, and "how do you know if there is a fire at your house?" This information should be presented in several short sessions rather than all at once.

Older children can be taught how to open windows for escape. It is extremely important that children be taught to go to a window during a fire situation.

The entire family must be involved in a home exit drill plan before it can be effective. The amount of material taught to a child is dependent on the child's age and capacity to retain information. Younger children will need more frequent practice for reinforcement.

Material to aid in teaching exit drills can be very inexpensive and does not necessarily need to be elaborate. The Golden Book "Bambi" can be used effectively to introduce the subject.

There are several storybooks on the market today which can be used to introduce the topic of exit drills in the home.

- "Bambi"

In the story of Bambi, a fawn has to cope with a fire in the forest. Bambi's mother dies in the fire. Read or tell the story and then use a discussion like this one:

Where does Bambi live?   The forest
Where do you live?   House, apartment, mobile home
What happened to Bambi's home?   Had a fire
What did Bambi do?   Got out of the forest
What would you do if there were a fire at your home?   Get out
Boys and girls should GET OUT

The amount of information conveyed at this point should vary with the age of the children. The older the children, the more information. With most, leaving the strong impression of exiting (GET OUT) with this story and following up with more information in another session is recommended.

- "Sam the Fire House Cat"

In this story Sam rescues Mrs. Catz and her daughter Becky from a fire at the warehouse across the street from the fire house. One of the Little Golden Books (#580), this title is available at stores across the country. It is a fun story with a happy ending that children enjoy. It also serves as a good introduction to emergency exits. Mrs. Catz and Becky are in the smoke and coughing. Sam leads them to safety. Children need to exit from a fire by crawling under the smoke.

Read the story to the children and review pages 13 and 16 through 19 which show the smoke. Ask the question: *What would you do if you were caught in smoke?* Crawl under the smoke to exit. Practice crawling under smoke.

Materials For Practicing Home Exit Procedures and Skills

Use several black crepe paper or fabric streamers, three to four yards long, or a dark gray sheet to simulate smoke. Have several children line up, forming two lines opposite one another, and hold the streamers or sheet "smoke". The children should alternate crawling and holding the "smoke" so everyone has an opportunity to do both.

Home exit skills can be practiced in a number of different ways. The skills which need to be practiced are: manipulating various types of door locks such as a chain lock and dead-bolt lock, opening or raising a window and then pushing out a screen, and getting out of a first- or second-story window properly. The latter would include sitting on window sill, laying across window sill with legs outside, hanging on outside the window and then dropping to the ground. In rooms above two stories, stay inside and open a window. An actual door and window are excellent aids to practice these skills. These are easily obtained or built by firefighters.

HAZARD RECOGNITION

Flame, scald and electrical burn hazards are identified as high risks to young children. The objective in this area of fire safety education is that children be able to identify these potential burn hazards in their everyday living environment.

Scald burns outrank all other burn injuries for young children. Flame and contact burns associated with hot surfaces also rank high, as do electrical burns resulting from children chewing on electrical cords.

When children can identify potential personal danger in a burn hazard situation, then they are better prepared to avoid the hazard and thus be responsible for their own behavior in association with these situations.

Teach children which is the hot water and which is the cold. Also adjust hot water heater thermostats to avoid scald burns.

Drinking hot beverages while holding a small child can result in scald burns. Some parents may not realize potential hazards like this and will take precautions once the hazard is brought to their attention.

Burn Hazards Identified

- Scald Burns: hot bath water, hot beverages, hot tap water, hot foods

- Contact Burns: open flame, hot cooking appliances, hot heating units, household appliances

- Electrical Burns: electrical cords, electric sockets, extension cords

This area of hazard recognition for young children can be greatly influenced by environmental modification. Suggestions to parents and others who care for young children are helpful in reducing the overall risk of the children. Keep the following considerations in mind at all times:

- Turn down thermostat of water heater (130 degrees).
- Carry food to table only after children are seated.
- Never hold children while drinking a hot beverage.
- Supervise children at all times.
- Be extra attentive during stress times such as breakfast.
- Keep children out of the meal preparation area.
- Store "goodies" other than over the stove.
- Install grills and protective gates around heaters.
- Use safety caps on electrical outlets.
- Avoid the use of extension cords.
- Keep electrical cords away from child's reach.
- Keep appliance cords on top of cabinets.

Turning down the thermostat of water heaters will lessen the probability of scald burns.

Carrying food to the table after children are seated eliminates the possibility of being bumped by a playing child and spilling hot food.

PUBLIC FIRE EDUCATION PLANNING — GRADES 1-3

It is felt that one of the best ways to reach children with Public Fire Education is through the school system. Within the schools, you have representatives from the majority of homes in the community. Therefore, you should emphasize that the children take home what they learn and share it with their families.

Working with Grades 1 through 3 proves difficult for some firefighters because they feel they cannot relate or "get down on their level." However, it is necessary to make younger children understand what you are talking about or you do not accomplish your goal. Talk with them in the same way you would talk to a small child anywhere in a friendly, caring manner. You might start by mentioning some things in the room or comment on how nice they look or how well behaved they are. Let them know that you care what happens to them and you do not want them to get burned in a fire.

After you have established a friendly air with the children, go ahead with your class, which should include more than just showing the fire apparatus.

Showing the fire engine is an important part of the overall program; however, it cannot take the place of fire education. In class you should explain to the children, on their terms, the importance of putting matches away, how to avoid getting burned and how to exit safely. Talk positively rather than in negative tones. You should not just say "Don't play with matches." Instead, instruct them on what to do with this useful but potentially dangerous item.

Stress positive actions and behavior concerning fire hazards

With lower grades, visual aids are a necessity. It need not be something fancy; a simple drawing or poster will work nicely. Anything to which the children can relate and react serves a useful purpose. Be especially careful in using films with this age group. Make sure that those used are not above the children in subject matter, language and general approach. If they don't understand it, the film can't help them. Before showing a film, explain to them what it is about and what you want them to look for. After they see it, emphasize important points conveyed in the film.

Emphasize what to do in case their clothing catches on fire: Stop, Drop and Roll. You can have one or more children demonstrate how this is done. One idea, if time and space allows, is to let the children get in a circle and throw a small ball to various students. When one catches it, he pretends that his clothes are on fire and acts accordingly. Also, you can use a small flame cut out of felt or paper to stick on individuals to practice the Stop, Drop and Roll procedure. (See pages 36 to 39 for additional details on these topics.)

Emphasize the Stop, Drop and Roll procedure

Another important topic to include with the younger children is "Exit Drill In The Home," otherwise known as Operation EDITH. Naturally, first- and second-graders cannot understand a complete plan as you would teach it to adults. However, there are various aspects of home evacuation that should be taught to the younger grades including the following:

- Have exit drills for each home.

- Stay low to the floor in smoke.

- Go to a window if door is not accessible.

- Stay at a designated meeting place once out of the house.

You can draw a sample floor plan showing the escape routes on poster paper and use it as you explain the important points of evacuation. Explain the floor plan as a map showing their home, just as if they were on top of the house looking down on it. On their floor plan, parents should mark each room showing whose room it is and showing two ways of getting out in case of fire — the regular door and one more way, usually a window.

If using a film on home evacuation, it is a good idea to show your home floor plan and explain emergency evacuation first. Then show the film and follow the film with an emphasis on the key points.

Talk about home fire hazards with this age group. Explain to them the dangers of some common hazards found in the home. Be careful not to get too technical in this discussion, but show examples or pictures of the hazards so they will know how to recognize them. Show the children a worn electrical cord and caution them never to touch it but immediately tell their parents.

One serious hazard which many in the fire service tend to forget (probably because it is not a direct fire hazard) is a pot or pan left in a position to be pulled off the stove. This situation can result in scald injuries to the child. We generally think of this regarding smaller children but it can easily happen to a six- or seven-year-old, especially if they are "helping Mom" in the kitchen. Therefore, include this in your program on fire hazards.

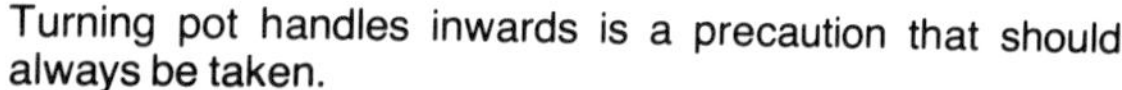

Badges, or some other small token, can be given to children to take home. This will help them to remember fire safety.

Turning pot handles inwards is a precaution that should always be taken.

Snuffy, the Burger King talking fire engine, discusses some of the finer points of fire safety with young admirers.

This program should include precautions about heaters, fireplaces, electrical appliances, lawn mowers, grills, trash fires and stoves. Naturally, some hazards should be stressed more at certain times of the year. For instance, you would talk more about lawn mowers in the spring and summer and discuss heaters in the fall and winter.

The attention span of a child in Grades 1 through 3 is relatively short; therefore, do not drag out your program but get to the point and finish in approximately 30 minutes. Give the students time to ask questions but do not let them talk on or about incidents which happened to them in the past.

After your class program is concluded, ask the students to look at the fire truck if you have planned it. When showing a fire truck, keep the explanations of equipment simple. Kids neither understand nor care how the pump works, for instance, but they enjoy seeing the hose, ladders, breathing equipment, protective clothing and small tools. If time permits, demonstrate how the hose throws water and let them take turns holding it (with your assistance).

Before leaving, hand out any useful fire prevention literature, preferably something like a coloring sheet for the children and a folder of material to be taken home for the parents. Invite them to visit the fire station.

PUBLIC FIRE EDUCATION PLANNING — GRADES 4-6

With a group in Grades 4 through 6, you will be talking about the same basic topics as with the lower grades, but your delivery will differ somewhat. You can go into more details, but still keep it so it can be understood. Technical explanations such as the way to operate the truck (pump, gauges, etc.) should be avoided. Some students will ask about them, but keep the answer simple.

After students reach the fifth or sixth grade, they are not quite as fascinated by the fire truck. Therefore, you may have some in the group who get bored or easily distracted, so be alert to those who may try to hinder you or the other students. Make it clear at the beginning that they should not climb on the truck or handle the equipment unless you give permission. Another firefighter should be present to help control the students. Most of the time you will probably be with your company so you will have assistance.

When there is more than one person showing the truck, you may want to break the class up into two or more groups, so that one starts on one side and another person starts on the other side. This way, there will be fewer students in the groups, making it easier for you to talk to them. This works especially well if you are showing two pieces of apparatus. One person starts a group on one truck and a second starts at the other truck.

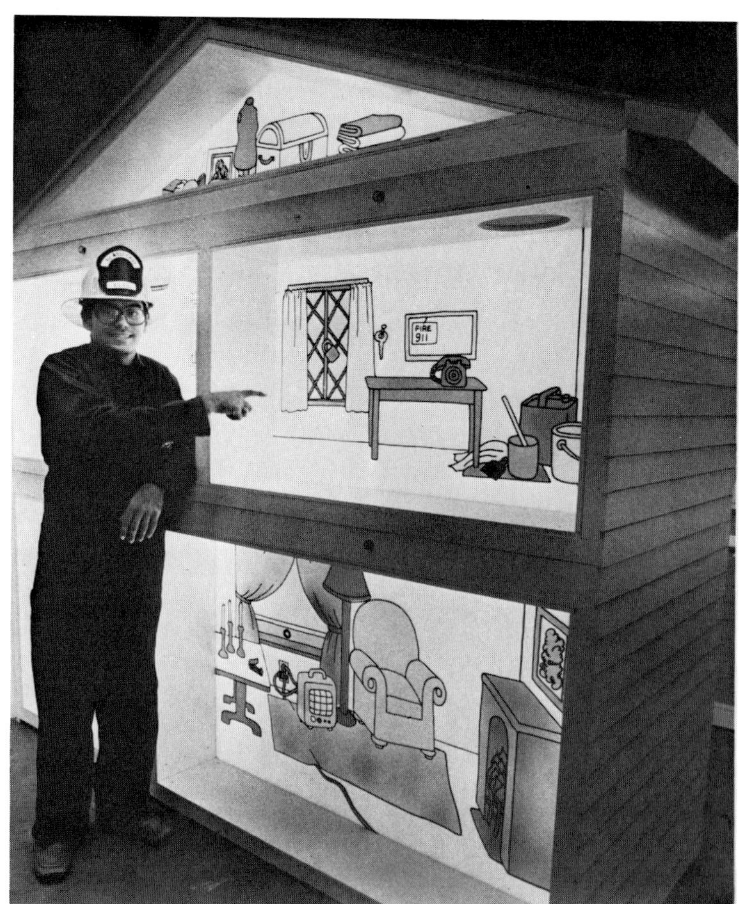

A "House Full of Hazards" is what Fireman Fred discovers as he discusses with children what they can do to eliminate fire danger in their homes. The house is part of the National Fire Safety Exhibit and Show touring shopping centers that is sponsored by Burger King Corporation.

With older children, home hazards can be discussed in more detail. Visual aids are a good method of showing common hazards which the children should be alert for in their homes.

In class, you must control the students at all times. Do not let your time be taken up by students telling stories which may not be to the point. The trouble is that this may not be known until listening through a too lengthy discourse.

Anything you can take to class to help you explain your topic would be very helpful. For instance, if you are discussing gasoline storage, take a safety can to show a proper storage container. Emphasize the importance of not playing with flammable liquids.

Electrical hazards can be explained in more detail with an explanation or demonstration of the hazards resulting from improper fuse sizes, poor electrical wiring, the use of electrical appliances around water, and climbing trees, poles and towers and touching high voltage lines. Serious or fatal injuries can occur as a result of playing around these. Electrical burns are discussed elsewhere in the manual. The danger of too many cords plugged into one outlet as well as cords run under rugs should also be mentioned. When discussing these, take a cord which fits the description to use as an example.

Students may have occasion to use the lawn mower, light the charcoal grill, build fires and use lanterns. Therefore, these topics should be covered thoroughly in the spring.

Encourage students to practice good clean-up habits at home and explain the value of good housekeeping. Relate an experience where your fire company responded to a house fire in which trash and debris added fuel to the fire. Also, it is important to clean up after special building projects, or other times when a certain amount of debris is generated.

Other special hazards should be mentioned such as problems at Christmas, Halloween and other special occasions.

If you are teaching home evacuation, you should go over the instructions for planning and conducting an exit drill. Also, consider their age and formulate specific rules on how to get out and what to do if trapped in a fire.

A film could be used to show why a plan is needed and how to make and use it. The film should be a good one which holds their attention and tells the facts without a lot of over-dramatization. Certain important points should be emphasized after the class views it. These would include drawing the floor plan, showing the emergency and normal exits, describing how to use it (practice it often, but not alone — only with the family), and explaining how to get out of the window by jumping if possible and, if not, by the dropping method. For houses with additional living levels or very high windows, point out that the best way to escape is with a folding escape ladder. Teach the students to roll out of bed and stay low to the floor, test the door and escape in the safest manner. Explain the importance of having a place to meet. Someone should be assigned to call the fire department from a neighbor's house. Also emphasize to them the necessity of never going back into a burning house. (Smoke detectors are covered in another section of the manual.)

Another subject which should be emphasized is clothing fires including how to prevent them, what to do when it happens and how to treat the burn injuries which are received. The burn injury is explained in Appendix B of the manual (See page 161).

Explain in detail that the person whose clothes are on fire should not panic. Running fans the flames and makes the fire burn faster. Practicing STOP, DROP AND ROLL should be emphasized, and whenever possible, have a student demonstrate for the class how this is done. Boys and girls in this age group are more apt to set their clothes on fire while playing with matches than any other age group. They should be encouraged to leave matches alone and to set a good example for younger children by putting matches away.

Because they often cook for themselves, students in the fifth and sixth grades are old enough to learn how to handle a small grease fire in the kitchen. Methods of extinguishment such as sliding the lid on or the use of baking soda should be explained.

FIRE EDUCATION PLANNING — JUNIOR AND SENIOR HIGH SCHOOL

In junior high and high school, the same basic information should be given as explained for younger students. The subject matter does not change, just the method in which it is delivered. Various styles of presentation can be used such as films, lectures, demonstrations and slide programs. Successful presentation has to begin when you walk into the class. Your mannerisms, attitude and speech can affect your overall program. In most cases it works very well for a younger firefighter to go into a junior or high school because he can relate to the students better and will probably have a little more patience with this age group. He can perhaps more easily "speak their language" and it may be better if he wears street clothes or department work uniform. A full dress uniform gives the law enforcement type of image and may "turn them off" at the beginning.

Films are good visual aids to use with high school groups

Relaxing with this age group is more important and can be accomplished by such things as moving about the room, sitting on a desk, talking briefly about cars or other interests they have. Also of importance is to convince the students that you care about their safety and want them to learn how to protect themselves. Tell them the facts realistically and don't use scare tactics. If you become overly dramatic, they will not believe you; and if you are dull, you will lose them. Therefore, be yourself, be truthful and be interesting.

Explain that such things as practicing an "Exit Drill In The Home" or the "Stop, Drop and Roll" for clothing fires is not just for kids and it isn't silly for young people and adults.

Use your experience as a firefighter to strengthen your case if necessary. Explain what it's like to see someone's house burn to the ground, or explain how it feels to drag a human body from a fire. Let them know how important it is to you that you don't have to do this, that you had rather be sure that they know how to prevent a fire or injury and that they know what to do in case it happens.

Express sincere concern for the public's safety

With the junior and senior high school age, you have the opportunity to work in various ways such as Boy Scout Firemanship Merit Badge Counselor and speaker in vocational exploration classes. With both of these classes (and others which could be included such as shop and home economics, etc.), you are able to reach students in different ways to let them know that your job is important. You can be just as helpful to the general public off the fireground as you are when fighting a fire.

THE YOUTHFUL FIRE-SETTER

Because the youthful fire-setter problem is usually very complex, it should not be lumped together with the routine arson cases. Some common causes of fire setting in young people include boredom, vandalism, peer pressure and curiosity. The problem may go far beyond the fire itself. It usually relates to physical or emotional problems and it needs to be dealt with carefully. Because the average firefighter has not had special training in this area, it is recommended that these problems be handled by those who are prepared to work with the fire-setters in cooperation with the Fire Prevention Bureau, local law enforcement and juvenile agencies. It is advised to learn what agency in your community would handle this type of problem so that it is possible to notify them when it arises.

ADULT FIRE EDUCATION

Adults are supposed to be the most responsible citizens yet being an adult does not automatically mean being fully informed about fire safety. Adults often hear only what they want to hear when it comes to fire education. These "responsible" citizens start and put out more fires than any other age group. They are burned seriously less often, but they are the relatives of younger and older people who are suffering unnecessary burns. Adults have the idea that fire can't happen to them; and we know that this attitude is the most dangerous thing of all because it puts up a barrier between you and the person. Therefore, we must work to change this attitude so the adult public will be responsive to the information you have to offer.

You can work to change the attitude by approaching civic clubs, school P.T.A. groups, church groups and any areas where adults meet. Contact the presidents or leaders of the various groups and express your desire to speak to the members on fire safety. Some will respond immediately and some will be hesitant for various reasons. If you can't make a definite arrangement to meet, follow up a few weeks later to let them know you are still interested in helping the people become more aware of the fire problem.

Once you have a meeting scheduled, be there on time (actually a little early is best so you can meet the leader and get your equipment set up). Most programs will run approximately

Youthful fire-setters usually require help from specialists. Know where the experts are located in your area for referencing problem cases.

20 - 30 minutes. Make sure you have everything you need such as projector, spare bulb, screen and drop cord. You may check ahead of time since some groups have their own equipment. Be careful in using a projector other than your own since you may not be familiar with its operation. Have a few biographical facts about yourself written down to give to the person who introduces you (how long on the department, etc.). Of course your personal appearance is very important.

As you start your talk, relax, smile and get your audience "with you". If you can add humor, do so if it is done tastefully and comfortably. Emphasize that you are there to help them become safer citizens. With adults you can point out such fire facts as:

- Most fires are residential.
- Three out of four household fires are put out by a member of the family, usually a woman.
- Thousands of lives are lost by fire and burn injury each year.
- Smoking is a major cause of fires and burns in the U.S.
- Three out of five fires involve the ignition of grease or other foods.

Adults will probably respond more quickly to local happenings, so take advantage of your experience as a firefighter and relate to fires in the area. Tell what happened, what could have been done to prevent the fire or accident, and also tell any positive side. If a smoke detector saved the family, for instance, make a big point to talk about it, especially if your talk is on home exit drills. Tell them about the ones who were saved as well as the less fortunate, because people like positive thoughts.

Adults will also respond quickly to your reference to children. Relate the information to them in a way that lets them know that teaching and practicing fire safety is a parent's responsibility and not just an option. Don't preach or lecture to them, because adults don't want someone coming in and telling them what to do and what not to do. Instead, let them know you care what happens to them and their families.

Smoking is a major cause of fires and burns in the United States.

You should appeal to the basic concerns such as where they live. You certainly would not talk to a rural garden club on how to get out of a high-rise apartment building. Adapt your material to what is most informative to that particular group. Also, remember seasonal problems so that you talk about portable heaters in the fall and winter and not in summer.

Some subjects to offer adults would include home exit drills, common fire hazards (especially electrical and flammable liquids), what to teach the children, clothing fires and burns, kitchen fires, using fire extinguishers and how to notify the fire department in an emergency. Emphasize that prevention is much better than having to deal with a fire after it occurs.

Adults must be made to realize that fire can strike in the home of anyone. Income, age, religion or race has no bearing whatsoever. Everyone is susceptible to fire.

Encourage questions and visits to the fire station, and let them know you are always there to help — not just when the fire bell rings.

Adults can be trained in specific prevention and suppression efforts from regular maintenance of smoke detectors to handling fire extinguishers. They should also be taught to keep their house free from common fire hazards.

FIRE EDUCATION FOR OLDER CITIZENS

In order for the firefighter to work successfully with older persons, it is necessary to understand them. Their goals and expectations are different but no less important than any other age group. They deserve extra attention because older people lose their lives and are seriously burned more frequently than younger adults. Therefore, we have a responsibility to this age group which has not been met in the past.

Many older people live in substandard conditions, have low incomes and have severe health problems, all of which can add to the fire safety problem. It is necessary for them to be educated in how to deal with their problems. For instance, the danger of smoking while consuming alcoholic drinks or taking medicines should be emphasized. This is the number one cause of fires and burns in this age group. Older adults often wear loose fitting clothes which, because they sometimes are not completely alert, easily catch fire around stoves and portable heaters. Proper use and care of heating and cooking equipment should be taught, as well as general safety in the kitchen.

Home evacuation for older citizens is extremely important and should be taught. Explanation should follow the same basic concepts as mentioned earlier for other groups; however, certain problems should be addressed. The importance of older persons sleeping on the ground floor and knowing how and being able to get out must be taught. If they cannot get out alone because of physical limitations, they should have someone close by on whom they can call for help. Also, they should learn how to notify the fire department. The fire department should be aware of those within their districts who are physically unable to escape so that the firefighters can prepare to handle the evacuation upon their arrival at the scene of a fire. Special decals are used in some areas to denote the location of such persons.

Many older citizens are happy to volunteer their services for fire prevention efforts. Solicit their cooperation, train them in the materials being used and allow them to spread the word.

When meeting with a group of older citizens, don't talk down to them but express your sincere concern for their well-being. Usually, a more relaxed atmosphere works especially well. In most cases, you should certainly relate to them as responsible adults and citizens realizing that they too want their surroundings safer. You can appeal to their desire to make their homes safer for visiting relatives and to serve as good examples to younger people. Let them know that you recognize their maturity and experience in living and that they can pass this on to others.

GENERAL SUGGESTIONS

1. Before you can successfully teach Fire Education, *YOU* must believe in Fire Education. You must have the sincere belief that Fire Education can make a difference in a person's life.

2. When talking Fire Education, draw on your experiences as a firefighter. No one knows better than you what lack of fire knowledge can mean to a person and their family and home.

3. Talk honestly about fire and its dangers but avoid "scare tactics."

4. Practice what you "preach." Let your audience know that you do the same things for safety that you want them to do.

5. Be on time and be prepared for meetings. If a question comes up which you can't answer, tell your audience you will find the answer and let them know. Don't give any answer just to look good.

6. Get familiar with visual aids and use them effectively. Learn to be innovative and make your own if what you need is not available.

7. Realize that Fire Prevention and Education should not be isolated to one week per year. Make it a continuing project throughout the year.

8. Learn the facts about the fire problem in your community as well as in the nation. Learn what causes the most fire deaths and injuries and attack that problem with a sound education program.

9. Investigate various ways to educate the public and use the ones most practical. In a small community, personal contact through schools, civic clubs and other groups may be best, while in large metropolitan areas, the media may be best.

3
Seasonal and Specific Group Programs

A special program might be almost any type of program. It should be directed toward the needs of your group or the problems of your community. There are programs presently available for almost any situation you may encounter in the fire service. There are all types of programs in many different forms (slides, slide cassette, 16mm and 8mm movies, filmstrips, video-tape, table top demonstrations, prepared lectures and many others). Special programs are limited only by the imagination and present an opportunity for you to be creative. Many very good programs are being developed by firefighters for their communities. Most of these have been specifically developed to meet a particular need. Some of these may be the result of a program that captured the attention of the firefighter, and its concepts or ideas were adapted to the fire service. Don't be discouraged if you don't have the technical know-how to do the entire program. Discuss your ideas with members of the department. They may have the knowledge or skills that you lack and may be able to help, or they may know someone else that can.

Be enthusiastic about the idea. Tell yourself and others that it will work. Your enthusiasm must be maintained throughout the entire production of the program. It is easy to become discouraged when occasional problems arise. When this happens step back and take a long look at the program, its ideals, its concepts, what it will do and how it will help your community. Keep in mind that if only one person is spared the pain of disaster by fire, then your program will have been a success.

Most of the public is not really aware of the potentially tragic force of fire. Individuals may not act to their best advantage during a fire situation since panic often results in irrational and non-productive behavior. It is up to the fire educator to not only tell individuals how they can prevent fires but also how to react if a fire occurs.

Living in a crisis-oriented society, most people do not even think about fire unless it happens close to them. Those who have been involved in a disastrous fire will usually be concerned enough to learn about fire safety, but others may not seek information. It is therefore your job to produce fire education programs and effectively disseminate information.

Evidence exists that fear or threats of violence, unless continually maintained, have very little long-term useful effect on a person's overall behavior. If information is contrary or disturbing to people's beliefs or inclinations, they tend to block it out of their mind, refusing to accept it. For this reason, public education on fire prevention and emergency actions should be presented in a positive way. If you are in charge of a program: KEEP IT POSITIVE! Most people feel that fire prevention and emergency action information should be in the form of short stories or a short dramatic example. Repetition is very important in conveying factual information. People want the facts. In all programs, emphasize information and repetition rather than impressions.

People may tend to block out negative approaches to fire safety such as scare tactics. It is much more effective to use a positive approach to public fire education.

POSTERS AND SIGNS

We are affected by posters and signs when we drive down the street of any city. Banks want us to borrow money from them. Insurance companies tell us that we are safe in their hands. Products of all types are advertised by billboards, posters and signs. Our homes are filled with posters of movie stars and plaques with short one or two line sayings. Why not have the same thing available to the public with a fire prevention message?

Graphically, a poster should be strong, wild or attractive and made of heavy weight construction paper. Posters that contain a check list of preventative measures can be taken home for reference and are extremely useful. These posters must be decorative enough that people will post them in different locations throughout the home. There will need to be a variety of subjects aimed at different groups of people.

- A "Sparky" or a cartoon poster for children: *"STOP! DROP! ROLL!"*

- A "funky" poster for high school and college students: *"PEOPLE CAUSE FIRES!"*

- A psychedelic poster for junior high and high school students: *"COOL IT! CIGARETTES CAN KILL"*

- A "plain" poster for adults: *"PLAIN FACTS ABOUT FIRE: DO'S AND DON'TS"*

Posters are limited only by the imagination and can be serious or far out. Posters may be placed in many areas: elevators, bus stops, cabs, subways, stores, school hallways — the possibilities are endless.

Posters are relatively easy to produce and can be used to convey seasonal fire safety messages. Managers or owners will usually be glad to display posters in their stores and, by selective distribution, a great number of people will be exposed to the message.

Not too many years ago there were signs placed along the roadways by Burma Shave with little jingles on them. Kids would look for them and enjoy trying to be the first to see and read them. If you have a small community, one-liner signs might be placed near the edge of town on the main street. Some suggestions might be: "Fire Hurts!", "People Cause Fires!", "Stop Fires — Save Lives!" These signs can be displayed on fence posts, telephone and light poles. (Be sure to get permission first.) It may be possible to get various civic organizations to sponsor a billboard about fire prevention in your community. Repetition is an important factor here. Even though people may not consciously think about fire prevention or emergency actions, a behavioral change may begin through constant and continuous exposure to the subject. We may subconsciously begin to remember those messages to which we have been repeatedly exposed.

Television and radio are mandated by law to donate a percentage of their time to public service announcements. Morning and mid-day fire prevention announcements might select women as the target audience with the message being fire prevention in the home. Saturday mornings and mid-afternoon announcements might be aimed at children to inform them how they can help in the prevention of fires and what to do if their clothing catches on fire. Evenings could go into the importance of the family working together on fire drills and establishing effective escape plans. Late night spots might discuss the problems of smoking in bed and the importance of early warning devices. Radio and TV spots should be presented with the announcements aimed at whatever age groups the regular programs are geared for. Interviews by the media on talk shows may be an excellent opportunity for special types of problems to be discussed.

Public Educational TV could present possible opportunities for both children and adult programs on the importance of public education on fire safety and prevention.

FIRE PREVENTION WEEK

Every year the week of October 9th is declared Fire Prevention Week. For some fire departments throughout the country this is the only time of the year that anything is done to promote fire prevention. During this one week the fire department usually does everything imaginable to tell the public about fire prevention. One of the problems is that at the end of the week the fire department has usually exhausted not only its manpower, but its ideas and someone usually says "Thank goodness, that's over for another year!" Nothing else may be done until the next year at Fire Prevention Week which leaves 51 weeks without any fire prevention activity. Those who are really interested in saving lives and property feel that the above idea is wrong. Fire Prevention Week should be the beginning of a 52-week campaign to do as much as possible with the people and resources available toward public education of fire prevention and emergency actions.

Fire Prevention Week should be a time when you inform the public of the many programs which are available to them. Through a planned systematic approach, the community problems concerning fire should be determined. Positive action to overcome these problems should be initiated. The following is an example of a possible outline:

List the various services your fire department offers. Note specialized equipment.

Divide the year into 26 two-week campaigns. This usually is enough time to get through a Special Program for different sit-

Determine who is most likely to be watching television during a specific time and target the message toward the audience. A late night spot might discuss the hazards of smoking in bed.

Permission is often granted to utilize signs and billboards to display fire prevention messages. Think about using this resource at other times of the year in addition to Fire Prevention Week.

uations. If not, extend this time period to cover the campaign properly. Use a calendar to note each holiday and plan a program to cover each of the holidays. Plan your campaign so that the program will begin at least two weeks before the holiday. You may need a follow-up program after the holiday.

Determine when school will start and end. What grades and ages are at each school? How can we best reach the children and by what method? Establish objectives and goals so you will know what you want to accomplish and then can measure the success of the program after its presentation.

Some goals might include:

- To teach children about fires — good and bad.
- What action to take in the event of a fire.
- How they can help prevent fires.
- Explain alarms and their importance.
- How to draw up a home escape plan.
- What to do if there is a fire in the home.
- Fire safety courses for babysitters.
- How to use a lid to extinguish fire in a pan.

Using a lid to extinguish fire in a skillet or pan seems obvious. Unless this procedure is emphasized, however, people may not automatically take this action should a fire occur.

Following are some suggestions on programs or presentations that could be used for school children, depending on the age group involved:

- Simple programs with one concept about fire safety.

- Art, reading and writing skills can be taught with fire safety as a subject.
- Station tours and visits to the classroom by fire personnel.
- Contests for fire prevention posters.
- Awards given to the children such as a ride on the fire truck, a picture in the local newspaper, posters adopted by the fire department for other use and ribbons given out to runners-up.
- Junior Fire Chief programs with certificates.
- Fire prevention literature distributed.
- Show movies and slides on fire safety.
- Collect newspaper articles and put them in a scrap book.
- Have book reports about fire safety or history of the fire service.
- Have the students write out questions about fire safety and then research the answers to their own questions.
- Let students discuss any fire hazards they know about.
- Set up discussions on different hazards and how to correct them.
- Creative artistic ideas for the student such as murals, showing the many jobs of a firefighter.
- Draw fire safety posters.
- Make a cartoon comic book for young children.
- Puppets depicting STOP! DROP! AND ROLL!
- Plan a skit for drama class.
- Have speech class write talks that can be given by them to others.
- Home economics classes can demonstrate to other students the proper methods of lighting a stove or oven.
- Science projects in controlled situations (how different chemicals act in fire situations, spontaneous ignition).

Some of the above suggestions could be given to the teacher for use as follow-up class work. Almost all areas of study can incorporate fire safety ideas and concepts but these must be planned with the school. As a fire department you will have a selling job to do to the school officials and a good working relationship must be maintained with them at all times. Your fire department may have to develop some of these programs but keep in mind that tomorrow's fire problems are here with us today. What the fire service does with our nation's youth will directly affect our future. We have a choice: we can either be part of the problem or part of the solution.

Start a Public Information Bureau where after dinner speakers can present lectures, demonstrations or slide programs to local civic clubs, groups and churches in your community on fire prevention and emergency action. Almost all clubs have a program chairperson who is responsible for planning club activities. Contact these people and let them know what is available. Be sure to tailor your programs for the group's special interest.

If possible, organize shopping center demonstrations, at least monthly. Advertise the day and times of demonstrations and have sufficient quantities of fire prevention literature to distribute.

Get involved in community affairs. Sponsor a little league team or benefits for the needy. Place a reserve engine out of service long enough to take the team on a parade through the community. Allow the station to be used for a meeting place for small community affairs. Assist in voter registration and provide the station for the polls on election day. Sponsor a scout troop or assist in sending children to camp. Provide an Explorer Post specializing in fire safety and fire fighting. Begin a cadet program to educate youths who wish to become future firefighters. Establish and be a part of community projects (clean up parks, restore historical landmarks, etc.). Use any and all opportunities to put fire prevention and emergency action ideas and concepts in front of the public. The more people they hear and see from the fire department, the more fire conscious they will become.

Provide special training for your community such as CPR classes, babysitting, safety classes, home inspections, what to do in the event of natural disasters, how to use fire extinguishers, industrial fire hazards training, and evacuation procedures from hospitals and nursing homes.

Home inspections are a valuable way to educate homeowners about fire hazards. If the program is adequately publicized, firefighters going into the homes will generally receive better cooperation.

By planning for the whole year, fire prevention activities will be well organized and they won't overlap. Anyone can also see at a glance how well rounded the activities are. Be certain that your plans are flexible enough for any unexpected problems. Just the knowledge of the programs' projected dates will allow personnel an opportunity to properly prepare for an upcoming campaign. It will also allow better use of personnel and money.

Since no program is really complete unless it is evaluated, keep good records. Keep good statistics and evaluate what effect the fire prevention efforts have had on the life loss, life injury and dollar loss in your community.

HOME FIRES

What are the causes of fire in the home? Why do they happen? How can one prevent home fires? What should a person do if there is a fire? Fires are going on continuously somewhere in America. What can we do to educate the public? What types of programs should we use to motivate the homeowner to learn about fire hazards and take the proper action to prevent them? We do not have definite answers for many of these questions but, for some of them, we do.

What Are The Preventable Causes?

Heating devices are often the cause of home fires and should be examined. Space heaters, electric heaters and fireplaces should be used with care. They should be placed so curtains, drapes, furniture or any other combustibles won't catch on fire. Don't overload the electrical circuits with heaters that require heavier wiring. Be sure to have furnaces cleaned and checked by qualified servicemen regularly. It might be desirable to put together some type of program just concerning heating hazards.

Fireplaces should be well protected and properly maintained and cleaned. If young children are in the home, the parents should educate them on the safe use of fireplaces. Allowing children to watch while lighting or cleaning the fireplace may satisfy some of their natural curiosity.

Cooking accidents are a major cause of home fires. Care should be taken to prevent grease build-up in the stove or range hood. When preparing food, don't leave the cooking area unattended. Know how to light a gas stove or oven properly. If it doesn't light by at least the second match, shut off the gas and wait a few minutes for the excess gas to clear. Have a lid for every pan or skillet that is in use. Learn how to put out fires by using the lids to shut off the oxygen to the fire. Be sure to immediately shut off the heat if there is a fire, and do not remove the lid until the pan or the skillet has cooled. While cooking, don't wear loose clothing, and be very careful not to reach across a burner at any time. Be careful in removing any pans from the stove — always use a hot pad. A program might be developed for use in school home economics classes or women's clubs and groups.

Provide large, safe ashtrays throughout the home and be **extremely careful** with smoking materials. Improper handling of smoking materials results in many residential fires. Never set ashtrays on the arms of sofas or chairs or any place where they may be knocked off. Be sure to put the cigarette in the ashtray so that it won't fall as it burns. Make sure that all ashes and butts are cold before emptying the ashtray into a trash container. Keep ashtrays emptied, and don't put tissues or other combustible materials in the ashtrays. NEVER smoke in bed! If you are prone to go to sleep in chairs or while relaxing on the sofa, don't smoke if you feel sleepy. Be very careful when re-fueling lighters so that you won't overfill them. Keep lighters and matches out of the reach of children. Before going to bed, always check to make sure all smoking materials are cold. Do not smoke near combustible materials.

Wiring should be large enough to carry the load for which it was designed. Check for loose connections, frayed cords, broken plugs, broken switches or outlets. If fuses or breakers are overloading, you should have a qualified electrical serviceman determine what the problem is since the situation needs immediate repair. Never overload a circuit. Excess heating of wires may cause a breakdown in the insulation or a fire in the insulation. If someone smells anything overheated, they should treat it as a potential fire. Disconnect the faulty equipment and call the fire department to investigate it. Only use extension cords when absolutely necessary. Make sure the size of the wire is not too small for the job for which it is intended. Do not run cords under carpets or throw rugs where they will be walked on. This wears the insulation off and can cause a short circuit and a fire. Never replace a blown fuse with a bigger one or short out the fuse.

A printed check list of home fire and burn hazards for distribution to adults can be effective. Warn of simple hazards such as reaching across hot burners and overloading circuits with extension cords as well as more complex hazards.

Gasoline, the most common flammable liquid, produces vapors that are heavier than air and will settle to low places. These vapors can travel invisibly along the ground or floor and be ignited by a spark or flame at a considerable distance from the source. Once ignited, those vapors can burn with explosive force. Never use flammable liquids near a flame source such as pilot lights. Even static electricity sparks produced by sliding across a car seat or walking across a carpet can produce a spark which will ignite flammable liquid vapors. Gasoline has been compared to dynamite. It has been determined that under the right conditions, one gallon of gasoline is equal to about 14 sticks of dynamite. Extreme care should be used when filling gasoline-powered equipment. Do not fill the engines if they are still running or hot. Provide ventilation when using flammable liquids because the vapors are dangerous.

Pressurized aerosol cans are commonly used to dispense paints, hair sprays, pesticides, wax, polish, antiseptics and many other household products. Many aerosol cans are like miniature LP gas tanks; explosions may be caused by high temperatures. Do not throw them into the trash where they might be burned or do not store them in heated areas or in the sun. Don't puncture the can, even if it appears empty, because it could explode. Propellants and contents may be flammable and react like a blow torch. Don't smoke while using aerosol cans. Some have toxic propellants or contents which if used without proper ventilation could cause dizziness, nausea, headaches and even death. Be sure and read the directions very carefully. We can, through public education, teach the homeowners what a hazard is, and what they can do to prevent a hazardous situation from becoming a disastrous fire.

This display board exhibits some of the major fire hazards that are found in homes. Emphasize the hazards of pressurized aerosol cans since many people may not be aware of the potential danger involved.

Why Do Home Fires Happen?

Carelessness causes fires. People just don't pay attention to the little details of safety. Some are negligent and reckless with things that they know are dangerous.

Ignorance causes fires. Some people simply don't know how to use fire safely. They have not been taught about fire hazards and what they can do to prevent a fire. Most importantly, they don't know what to do in an emergency which involves a fire. Ignorance concerning fire safety is one problem the fire service can do a great deal to overcome. Through public education, ignorance can be overcome.

Apathy causes fires. Apathy is nothing but an attitude of indifference. "I know, but I don't care." This problem is a very difficult problem to deal with. In order to change an attitude, the person must be made aware that there is a need to change. They must be made aware that fire can maim or kill them and the ones they care about. Each one must understand their responsibility to others as well as to themselves. It is not enough to know about the danger of fire, one must also care.

Arson causes fires. Arson, or fires which are intentionally set, is an ever-growing problem nationwide. Although this type of fire problem is handled by the Fire Prevention Bureau or state or local law enforcement agencies, the firefighter should be aware of his role in it. Programs are presented in many areas in an effort to combat this problem. You can certainly include information such as local statistics in fire education programs.

Facts without acts are failures. It must be remembered that we can only save that which is not already lost.

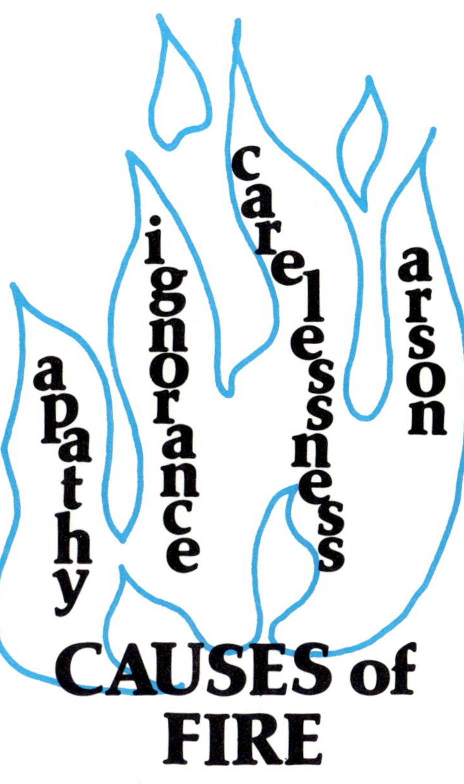

People's ignorance concerning fire hazards can be successfully overcome through the public education process. Apathy in people is an extremely difficult problem to deal with since correction involves attitude changes.

What To Do

If you detect even the slightest hint of a fire, alert others in the house. Scream, pound on the walls or blow a whistle. Stay low to the floor. Cover your mouth and nose with a damp cloth if possible because this will help filter out some of the smoke particles. Don't assume that you are safe just because the air looks clear. Carbon monoxide is colorless and odorless. Carbon monoxide affects judgment and may cause headaches or nausea. Before opening a door to another room or hall, be sure to feel the door as high up on it as possible. If it feels hot, don't open that door. Use another means of escape. One breath of super-heated air or flames could cause the air passages to swell and close off, causing death. If unable to escape from the building, go to the window and open it a little from the top and the bottom. This allows the smoke and heat to escape at the top and fresh air to enter at the bottom. Breathe the fresh air and await rescue. A bedsheet, blanket or article of clothing should be hung out the window to attract attention. If smoke is coming in under the door, stuff bedding or clothing around the crack. Don't try to hide under the bed or in a closet; stay by the window. Yell and scream for help. If above the second floor level, don't jump out of the window.

PROPER MEANS OF REPORTING A FIRE

The amount of time it takes to report a fire can mean the difference between life and death. The following information is needed:

- Get out of the burning building and call from a nearby phone.
- Know the right telephone number for the fire department.
- Give your name and the location of the fire.
- If in a large metro area, give the city.
- Phone number for call back purposes should be given.
- Do not hang up until dispatcher instructs you to do so or he hangs up the phone.

Special programs of all types may be drawn from this information from how to call the fire department to what are the causes of fire in the home. This information is all public education and it is the fire department's responsibility to make sure the residents in their community know the proper responses.

MOBILE HOMES

The hazards in a mobile home are not that much different than those in any other home, although mobile home construction may contribute to rapid burn rate once a fire is started. There are a few things that should be discussed briefly. Mobile homes should have at least two exterior doors that are remote from each other and easily accessible. This allows for a better chance of escape. Newer mobile homes must comply with American National Standards Institute A 1191 and National Fire Protection Association Standard 501B. Currently, mobile homes must have windows and doors to provide for emergency escape. Every bedroom must be provided with an exit window that opens

IN CASE OF FIRE
DIAL
Hampshire **683-2121**
REPORT
FARM No. ________
HAMPSHIRE FIRE PROTECTION DIST.

Printing and distributing information on how to telephone the fire department, including the correct number to use, is an asset to the public as well as to the department.

The construction of mobile homes produces rapid fire spread. Detectors are now required by law to be installed in newly constructed mobile homes.

easily from the inside. At least one smoke detector is now required. Mobile homes built before 1975 were not required to have smoke detectors or escape windows. The same fire prevention and emergency actions apply here as they do in home fires.

APARTMENTS

The hazards in an apartment are about the same as in home fires. Still, there are points which relate to apartments that should be discussed briefly.

- Know your apartment building. Learn the layout of the building, and most important, where the exits are located.

- Pre-plan your escape routes in the event of a fire.

- Know the location of fire extinguishers and how to use them.

- If the fire is in your apartment, leave it. Close the door, sound the alarm to let others know there is a fire, proceed to the exits and on to the outside. Notify the fire department. Know the location of your closest alarm box and what kind of alarm it provides (alarms in buildings may notify tenants only). Identify yourself to the arriving firefighters, and tell them the location of the fire.

- Never return to the building until the fire department determines that it is safe to do so.

- Once outside, go to a place of safety away from the doorways so the fire department isn't hampered by people standing around.

- If needed, special evacuation plans may be developed by the fire department and the apartment managers.

- Don't use elevators in fire conditions; use the stairways.

- Never assume that someone else called the fire department, do so yourself.

Apartment complexes may have special needs in relation to such things as escape planning. The fire department should consult with apartment managers/owners and assist them in developing special evacuation plans if desired.

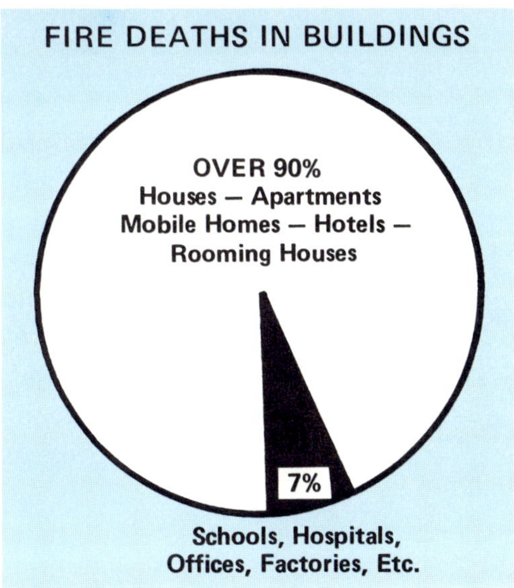

VOLUNTEER PROGRAMS

Anyone who has any special ability has a place in fire education whether the person is paid for his service or simply performs the tasks voluntarily.

Volunteer firefighters have put together some very good public education programs. They have spent weekends doing home inspections, shopping center displays, demonstrations, school programs and club talks.

It takes people, time and materials to put ideas into action. If the fire service can provide the ideas and materials, people will provide their time and talents. Ladies' fire department auxiliaries have learned fire prevention and home inspection techniques and have given many hours of service as volunteers.

Scouts have volunteered to help pass out fire prevention literature, clean up fire hazards in public places as well as many other things to help support their fire department. People want to help so get them involved. Usually all it takes is someone to provide some direction. Give people an opportunity to assist in public education.

The more volunteers that are enlisted to promote fire education, the more effective the programs will be. Volunteers can deliver presentations, help produce the programs or originate ideas for designing various programs.

Whatever people's interests, be it writing, art, photography, or public speaking, they have a place in public education. Maybe a person is very good at relating to preschool children; if so, they could put together a story about fire safety and teach young children. If they cannot initiate the program, provide them with the material and allow them to actually present it. Another person might be a very good sales person, so they could "sell" the idea of fire safety. Using volunteers for public fire education is extremely important. The more people you have working on a program the more people you can reach.

Some of the main ingredients of public education programs are sincerity, dedication, truth and imagination. Saving lives and property is the goal of public education.

FIRE STATION TOURS

Often the fire service has the opportunity to have children come into the station for tours. What is done with this opportunity is up to the teachers and the fire personnel. Many times the only thing that is accomplished is an explanation of the vehicles in terms that are unfamiliar to the children.

Tours should begin with an understanding by both teacher and fire personnel as to what the tour should be in terms of content and time. The tour should be well planned instead of just happening. Pre-arrange schedules so they won't conflict with normal routines or disrupt special activities. Always attempt to start and end on time so other schedules can be kept. Determine the age and interest of the group. Children of various ages must be treated differently. Determine the purpose of the tour. This can be accomplished by talking with the teacher prior to the visit. Find out what they think the tour should accomplish and discuss both views, theirs and yours. Once the objectives have been set, plan the tour so they can be met.

Have designated personnel with prior knowledge available to handle the tour. If the tour is *your* responsibility, cover the following points:

- Introduce yourself and your assistants.
- Welcome your group to your station.
- Explain how your fire department operates, including what to do if an alarm sounds during the tour.
- Show a general film of hazards in the home.
- Explain how, by correcting hazards, fires can be prevented.
- Explain what to do in the event of a fire.
- Don't attempt to show too much in one visit.
- If time permits, answer any questions or let the children tell you some story about a fire.
- Have literature suitable to the child's age to be given out as they leave.
- Ask the teacher, if possible, to follow up on explaining literature to the children after returning to school.
- Keep your tour content at a level they can understand. Never go below their level of intelligence, but don't talk over their heads.
- Communicate with them and make fire safety a pleasurable learning experience.

- Evaluate your program by asking questions to determine what they have learned.

- Always use a positive approach. Tell them what to do, not what they shouldn't do.

- Explain the WHY and HOW of fire safety.

- Keep a record of the tours, with such information as: How many? Children's ages? Date? Time? What subject was discussed? What literature was used? People to contact in the future?

- Be sure to invite the children back.

Children enjoy looking at fire engines and even handling the hose. When allowing children to participate in this type activity, be extremely careful and alert so that the children are not exposed to any dangers. In addition to these activities, fire station tours should include basic home fire prevention education.

CHRISTMAS

Christmas is a wonderful time of the year when people are usually happy. From joy can come sorrow if people fail to make fire prevention a part of their holiday. Educate the public for special fire prevention during the holiday season. Some of the items listed below would make good 10 - 15 second spot announcements on radio and television at Christmas time. Schools could send a list home with the children.

- Use a ball and burlap (living) tree for moisture retention.

- If you buy a real tree make sure that it is fresh.

- Leave it outside until you are ready to decorate it.

- Make a fresh cut in the trunk to facilitate water movement through the cells.

- Add water to it daily to keep moisture in the tree.

- Keep it in a cool place away from a heat register, fireplace or radiator.

- Keep it out of the traffic area and away from stairways and doors.

- Tie or otherwise secure the tree so it can't be knocked over by pets or small children.

- Do not smoke near the tree.

- Before using lights on the tree, make sure they are UL listed. Never use more than three strings of lights on one circuit.

- Make sure connections are tight and in good shape to prevent wire hangers and metal icicles from causing an electrical short.

- Never put lights on a metal tree.

- Take a live tree down as soon as it starts getting dry (when needles begin to fall). Don't take a chance.

- Make sure there are no loose bulbs or frayed wires.

- Be able to turn off the Christmas tree lights without reaching or crawling under the tree.

- Do not leave the lights on if you leave the house or when going to bed.

- If you have an artificial tree, be sure it is flame resistant.

- Make sure that all decorations are flame resistant.

- Dispose of papers and boxes at once. Do not allow them to accumulate.

- Don't burn papers and boxes in the fireplace.

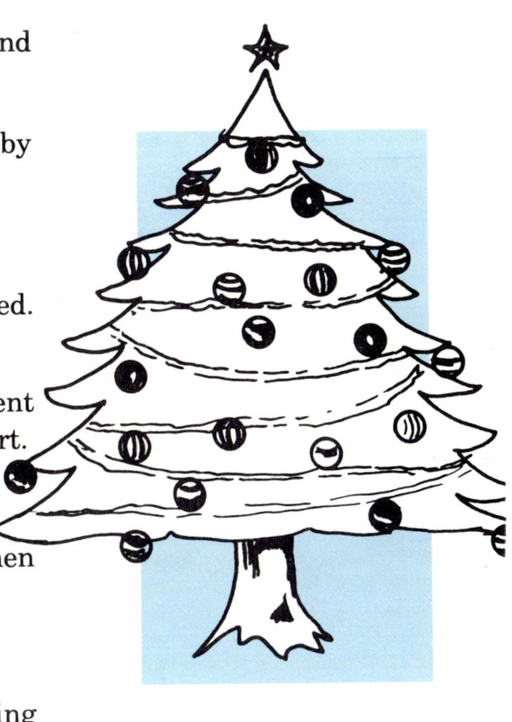

Tragic Fires Mar Christmas Holiday

BY THE ASSOCIATED PRESS

Christians celebrated the birth of Jesus and Jews marked the Hanukkah Festival of Lights today as the nation paused for gift-giving and prayer.

But there also was tragedy as fires killed at least 23 persons during the holiday weekend, including eight children of one family in New Orleans.

Much of the nation had a wet holiday, with snow in the Rockies, throughout New England and in parts of the Midwest and heavy rains along the Atlantic Coast.

President Carter and his family planned to spend a quite Christmas Day in his hometown of Plains, Ga.

Carter was to visit first with his mother-in-law, "Miss Allie" Smith and then his mother, "Miss Lillian."

Mrs. Smith was released Sunday — her 73rd birthday — from the hospital where she had been undergoing treatment for a

and handing out Christmas cards.

The group entered the base last May, and one of the demonstrators, John Calambokidis of Olympia, Wash., said they wanted to do it again at Christmastime because "it is a time of talk of peace on earth and goodwill toward people."

At least 27 persons died in Christmas weekend fires, including eight children in one family, and two children left unattended by their parents, authorities said.

One fire victim died while cooking a Christmas turkey. Three of the fatal blazes were attributed to faulty wiring on Christmas lights.

Burning crude oil in a storage tank forced about 2,000 people in the Baton Rouge suburb of Scotlandville, La., to spend Christmas Day with friends and relatives or at an emergency shelter.

The 3.3-million gallon tank, which was filled to the brim, collapsed late Sunday

Apartments, authorities said.

A Dallas family of four was killed in another Christmas morning blaze. The charred bodies of the parents and an infant were found huddled near the front door of their woodframe home. Investigators said it appeared they almost made it to safety. An older child found in his bedroom died later from smoke inhalation. Officials cited faulty Christmas tree lights as the probable cause of the fire.

The fire that killed the eight members of one family in New Orleans on Christmas Eve apparently was caused by a short-circuit in a string of Christmas tree lights, officials said.

The dead ranged in age from 1 to 18. Five of them were children of Ida Perkins and three were her grandchildren. She escaped with two other children.

In Indianapolis, authorities said two girls, ages 3 and 7, died from smoke inhalation in a fire that may have started

Newspaper headlines relate that Christmas is not always a joyous occasion for all people.

HALLOWEEN

Halloween is or should be a fun time for children. It is the responsibility of the parent to see that it is a fun time.

Clothing can and does burn, so be sure to keep children away from fireplaces, stoves, cigarettes, cigarette lighters, matches and candles. Be very careful of clothing fires since most costumes are loose fitting. When purchasing costumes make sure they are flame resistant, including masks, wigs and all accessories.

Never use candles to light up pumpkins or other decorations. Autumn leaves, corn stalks and berry branches, when used for fall decorations, may be a very dangerous hazard and should be kept away from heat sources, smokers, light bulbs and other possible ignition sources.

Bonfires, weiner roasts and other open fires should be supervised at all times and completely extinguished when the gathering is over.

JULY 4th

Fireworks can be fun, or they can maim or kill. Fingers, eyes and feet have been blown off with fireworks. Serious clothing fires have been caused by fireworks. Fireworks are not toys for children. Even sparklers can cause disastrous fires if used carelessly. Sparklers contain metal fragments which burn hot and fly off, possibly causing severe injury. If for any reason the clothing should catch on fire: STOP! DROP! AND ROLL! In some states the use of fireworks is illegal.

THANKSGIVING

Many disastrous home fires occur during the preparation of a special meal. Cooking, clothing, oven, grease, electrical appliance, careless smoking and many other fires happen during this time of year. Possibly visitors who are not careful or don't care can cause one mistake — the one disastrous mistake. Don't let this potentially happy and thankful time become a disaster. Extra care must be taken. Spot television and radio announcements along with news releases are a must at this time of the year.

SPRING CLEAN UP

Spring is when the homeowner disposes of the winter's accumulations. Paint, cleaning materials, aerosols and flammable liquids are common hazards at this time of the year. Once again spot announcements, special literature and news releases are a must.

NURSING HOMES

Thousands of people, for one reason or another, are living in nursing homes. These include the elderly and the mentally or physically handicapped. Most facilities are required by law to meet special fire codes, such as smoke detectors and sprinkler protection and fire-resistive construction. Many operators have all of this protection and feel very secure, thinking fire could not happen in their nursing home. These protection devices do not necessarily in themselves guarantee safe practices in other areas of a nursing home. The fact is three to four thousand times each year fires break out in these facilities, many of these resulting in multiple deaths.

Each employee should receive training in recognizing fire hazards

Not all facilities have this protection. What about the personnel of these places? Do they know about fire safety, prevention, evacuation, using extinguishers and proper housekeeping? The facility may have built-in protection, but what do the personnel do when there is a fire?

In one case, a small nursing home for handicapped children experienced a fire alarm at 2:30 a.m. when only four nurses were on duty. Having never been taught any emergency procedures, they became so frightened they ran out of the building. Since the panic hardware on the doors was set for night security, the nurses were locked outside, leaving 20 children still asleep inside the building with the fire alarm sounding. One of the nurses ran next door to another facility and called the fire department. Firefighters were forced to break into the building. Some clothes in a dryer were too hot, causing the alarm to activate; there was no fire. The next morning the fire department began a program to educate nursing home personnel about fire prevention, fire safety and evacuation procedures.

Fire department public education personnel should study the special needs of nursing homes in their community and teach personnel fire safety practices. Require each employee to spend some time learning what a fire hazard is, how it can be removed, how to use extinguishers, and proper evacuation procedures. This also gives the fire department an opportunity to find out any special hazards or problems that may be encountered in the event of a fire. There are many publications which deal with this special problem. The following are suggestions for improving life safety in nursing homes:

Train employees to use extinguishers and to properly evacuate residents

- Go to the nursing home and explain to supervisors why you are concerned.

- Set up a schedule for meetings.

- Tour the facility.

- Discuss any problems they may have.
- Attempt to show them solutions.
- Educate personnel.
- Follow up with regularity.

CHURCHES

Churches are often damaged by fire, with losses in the thousands of dollars. As an educator of the public, churches should be one of your concerns. The ages of the persons within the churches are from infants to old age. Fire personnel may be alarmed to discover that the nursery or children's classes are in the basement near the furnace room and storage areas where most of the fires begin.

Help your churches plan escape routes. Someone must be responsible for evacuation of the children and the elderly. One or two women cannot carry ten infants up stairs and outside. Panic may cause mothers and fathers to rush down steps to the basement after their children, making exit to the outside impossible. Young adults should be close to the older class to help those who need it. In an evacuation, have teachers take their attendance books so there can be a head count. Make someone responsible for notifying the fire department. Pre-determine some type of signal or alarm so everyone will know there is a fire. Practice this plan regularly.

Most churches look beautiful but may not be fire safe. You as a firefighter are responsible for the fire safety of your churches. Some of the common fire hazards in churches include:

- Electrical wiring
- Heating
- Cooking equipment
- Candles
- Housekeeping

Talk with trustees and church officials and show them what can be done to improve fire safety.

Explain the value of fire prevention, detection and escape planning. You may want to develop a package program for churches, one that they can show at their convenience.

FIRE SAFETY IN RECREATIONAL ENVIRONMENTS

In recent years there have been many changes in the expectations of people and in the general work and financial situation that exists in the United States. Workers continually press for and obtain a shorter work week and increased salaries. As less and less time is spent actually earning a living, more time becomes available to follow other pursuits. In spite of the effects of inflation there also seems to be more money available for recreation and entertainment. Recreation is now a very large industry which has considerable effect on almost every community. The economy of many communities is based almost entirely on the recreation and entertainment industries. A preoccupation with recreational pursuits should not result in forgetting fire safety.

There are many types of recreation environments. However, this section will deal with tents, tent trailers, campers, vacation trailers, motor homes and the great outdoors.

Tents

These structures have been in use for many years and there is no indication that they are going out of style. In fact there is an ever-increasing variety of sizes and shapes of tents being marketed. New materials have also been introduced to reduce the weight and bulkiness of tents to make them more suitable for hikers, mountain climbers and campers.

The newer fabrics used are generally fire retardant, but still may burn somewhat. Older tents are notorious for their rapid fire spread upon ignition. Retardant treatments used on fabrics have a tendency to lose effectiveness with exposure to the elements.

Water-repellent treatments also tend to lose their effectiveness and must occasionally be reapplied. Some water-repellent treatments make tents extremely flammable until they have been cured properly and have been completely aired out.

Tents are used as a home away from home. You can, therefore, anticipate that the same kind of activities that take place in the home (that most hazardous place) will also take place in or near a tent. People need a place to sleep, they must eat (which often means cooking), in many areas they need heat to keep warm (either with additional clothing and bedding or heat producing appliances). If they are in the habit of smoking or drinking, there is every reason to believe they will continue to do so when living in a tent.

There are many precautions that can be taken, depending on the activities that are taking place in or around tents. Some general precautions are as follows:

- Always assure that children are attended.

- Insure that matches and lighters are not left where children can play with them.

- Be extremely careful with cooking or heating appliances and with lamps or lanterns which could ignite combustibles.

- Refill flammable liquid appliances away from the tent, and do not store flammable liquid containers inside.

- Build open fires and place fuel-burning appliances far enough away from tent to prevent ignition from sparks, flames or heat.

- Do not use a charcoal burning appliance inside a tent.

- Do not smoke inside the tent, particularly at bed time. Careless smoking, particularly combined with the use of alcohol, is a deadly combination anywhere.

Tent Trailers

Tent trailers have been developed to provide a few more luxuries than tents. Most include some type of electrical system for lights. Ice boxes, refrigerators, stoves, and heaters or furnaces are not uncommon. With each additional appliance there is an increased chance of having a fire-related accident.

The general precautions referred to for tents apply equally to tent trailers. A few other precautions are also necessary. Insure that the correct type of propane tank is provided and is securely and properly mounted. Follow the manufacturer's directions for lighting, operating and maintaining various fuel-fired appliances such as furnaces, heaters, stoves and refrigerators. In most cases this means not traveling with propane-fired refrigerators and furnaces operating. Where stoves are provided, all the precautions used in the home should also be applied, including the provision of a portable fire extinguisher and lids large enough to cover cooking utensils. Take care not to place combustibles on or near heaters and furnaces.

Campers and Vacation Trailers

The last quarter century has seen the development of numerous makes, models and styles of both campers and vacation trailers. Many of these exceed the space and luxury of the earlier versions of the mobile home. The majority of them are self-contained units. They therefore include a variety of appliances similar to those found in mobile homes and other residences. Like mobile homes, campers and vacation trailers are of lightweight construction. The walls are thin, with insulation and vapor barrier sandwiched between the exterior skin and the interior wall surface which is usually a pre-finished veneer or hardboard. In newer models the interior flamespread characteristics are acceptable; however, the material offers little or no fire resistance. The compact size and good insulation lead to a very rapid buildup of smoke and heat resulting in an early flashover and complete destruction in a matter of minutes. For this reason it is imperative that extreme care be taken to prevent fires from occurring. It also would be wise to install and maintain early-warning devices to alert sleeping occupants quickly if a fire occurs, allowing them time to escape.

Install early warning devices in campers and vacation trailers

All the previously mentioned precautions should be taken plus any others that are covered under the headings for mobile homes and other residential occupancies.

Motor Homes

Motor homes are a relatively recent development. They simply combine the living accommodation and automotive chassis as one complete unit. The facilities provided are similar to those in campers and vacation trailers. Additionally, caution should be exercised in carrying out the chassis maintenance as well. The exhaust system should be kept in good repair to prevent carbon monoxide from entering the unit. Care should be exercised in refueling operations, maintenance and use of all accessories and appliances.

What is the role of the fire department in promoting fire prevention safety for this particular group of people?

As with any other field of fire prevention, fire department members can be very knowledgeable about the fire hazards and how to correct them, but unless they pass this information along to others, no progress will be made. The variety of ways to get this information across to the right people is limited only by the amount of initiative and imagination displayed by the fire department personnel. In many areas of the nation, vacation vehicle safety check clinics are conducted annually. Various other groups typically are organized to help in the checking for safety hazards. Participants may be mechanics or electricians — why not fire department personnel?

Talk to motor home owners' clubs about fire safety

To whom should you talk? Recreational vehicles are used to transport and house people of all ages from new-born babies to aged pensioners. Obviously, children must be old enough to understand what they are being told. This leaves the field wide open from kindergarten up.

Where do you find groups of people to talk to? Talk to people at schools, youth groups, service and social clubs, church groups, camper clubs, trailer clubs and campgrounds.

How can you get the necessary information distributed? By arranging to speak to any age groups mentioned you can disseminate your information. Literature can be distributed at places frequented by travelers such as tourist information booths, service stations, propane refilling stations and recreation equipment dealers. There is a great variety of posters and pamphlets available from the National Fire Protection Association and state and provincial agencies. For special problems not covered, develop your own material and have it printed.

GRASS AND FOREST FIRE HAZARDS

In many parts of the country long dry seasons can be expected. Grass and forest fires are often caused by the carelessness of people. Special announcements at these times, warning the public of the potential danger and the special education of people who live in these areas, are a must. This special education may be on a one-to-one basis with the home owners in these communities. Seasonal programs must be done on a community level with the special needs of the particular hazards of the area being taken into consideration.

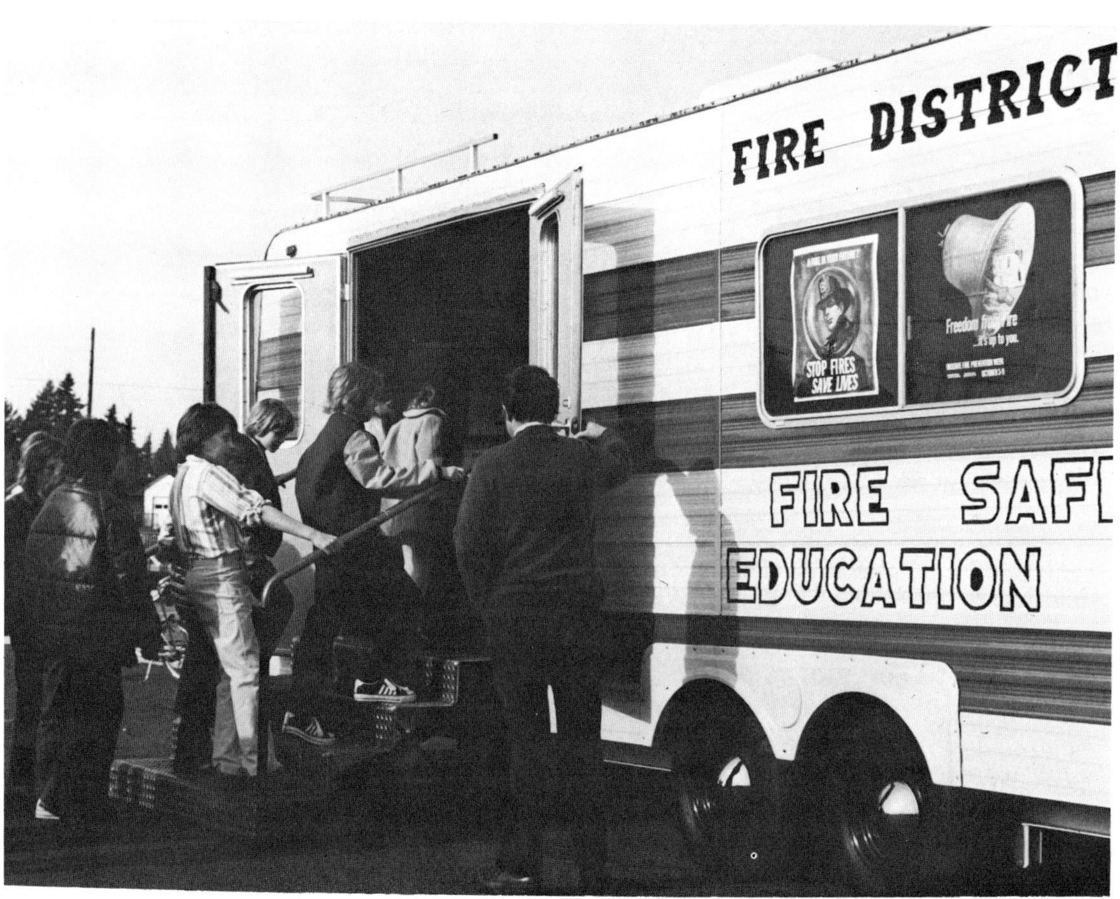

4

Smoke Detectors

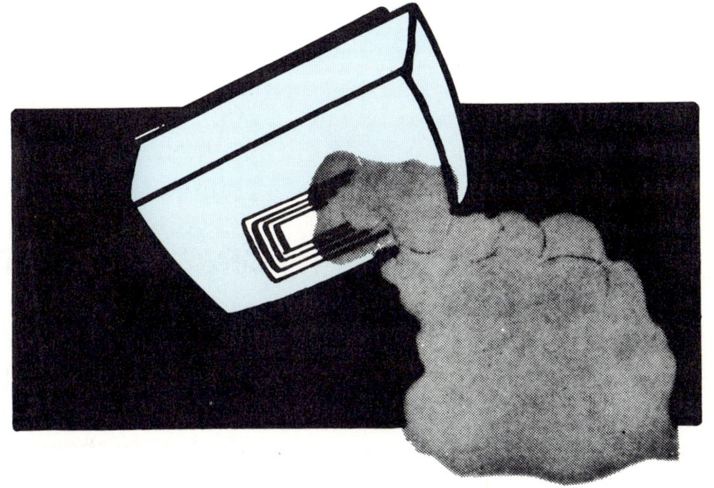

Smoke detectors save lives. Increasingly, communities across the country are reporting families saved from certain tragedy by the early warning provided by residential smoke detectors. This continuing evidence of effectiveness points out an opportunity for dramatically reducing the 6,600 residential fire deaths that occur every year in the United States. In fact, if all of our nation's homes were protected with smoke detectors, the residential life loss could be reduced by over 40 percent. This would have significant impact on our national fire problem, because over 90 percent of the nation's building fire deaths occur in residential buildings. The fires in offices, schools, hospitals and places of assembly get a good deal of attention, yet the major problem is the home fires that claim one or two lives at a time but with greater frequency.

Smoke detectors help the home occupants by giving early warning so they can escape. Also, the earlier the fire is discovered, the less property the fire can destroy before it is extinguished. In two very important ways, the smoke detector can also help the fire department. The alarm can warn of fire while it is still small, making extinguishment easier. Also, when the dwelling occupants escape from a burning home because of an early warning, the firefighter does not have to attempt an unnecessary or risky rescue.

Written by Dennis Ozment, Captain, Minneapolis Fire Department

Material from United States Fire Administration, Smoke Detector Technology Manual, Smoke Detector Training Manual, Smoke Detector Resource Guide

Fire fighting is the most hazardous occupation in the United States, and search and rescue is one of the most dangerous services the firefighter performs. While the early warning is often accredited with saving the lives of the home's occupants, let's not forget that the smoke detector's warning may have also protected a firefighter. When smoke detectors are not in the home, the firefighter is then forced to try to make up the difference.

TWO TYPES OF SMOKE DETECTORS

Available smoke detectors operate on one of two basic principles: Ionization and photoelectric. For maximum protection, the user should understand the advantages and disadvantages of both types.

Ionization Type

The ionization detector uses a small amount of radioactive material to make the air within a sensing chamber conduct electricity. When very small smoke particles enter the sensing chamber, they interfere with the conduction of electricity, reducing the current and triggering the alarm. The particles to which the detector responds are often smaller than can be seen with the human eye. Since the greatest number of these invisible particles are produced by flaming fires, ionization detectors respond slightly faster to open flaming fires than do photoelectric detectors. The radiation source in ionization detectors is not a hazard to the home's occupants. The U.S. Nuclear Regulatory Commission (NRC) performs a radiation safety analysis to determine that detectors meet safety requirements.

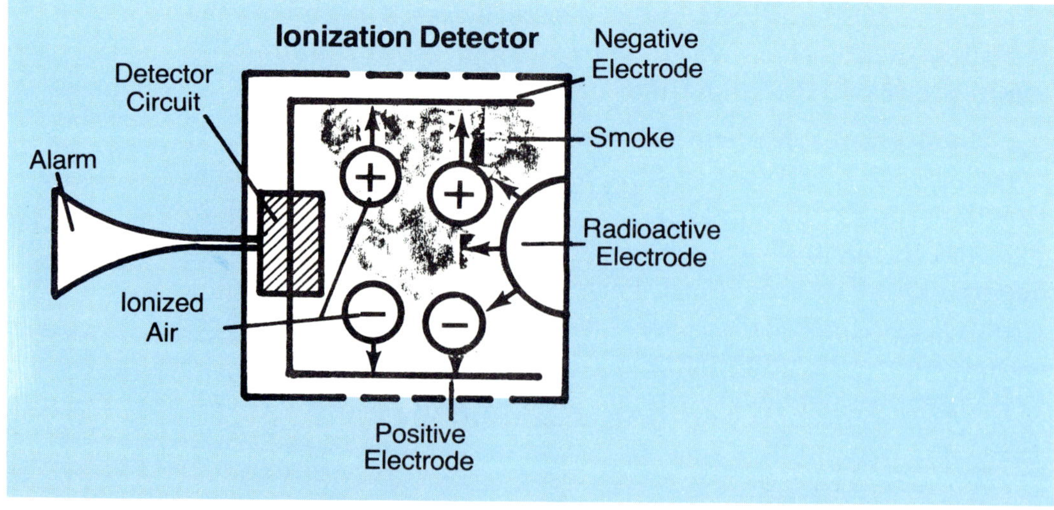

Ionization detectors are somewhat faster in detecting flaming fires which may produce little, if any, visible smoke.

Photoelectric Type

The photoelectric detector uses a small light source — either an incandescent bulb or a light emitting diode (LED) — which shines its light into a dark sensing chamber. The sensing chamber also contains an electrical, light-sensitive component known as a photocell. The light source and photocell are arranged so that light from the source does not normally strike the photocell. When smoke particles enter the sensing chamber of the photoelectric detector, the light is reflected off the surface of the smoke particle, allowing it to strike the photocell and increase the voltage from the photocell. (This reflection of light is the same means by which we see smoke in the air. That is, light from the room strikes the smoke and reflects it to our eyes.) When the voltage reaches a pre-determined level, the detector alarms. Smoke particles that scatter visible light are larger in diameter than those which an ionization detector senses. Since smoldering fires produce these larger smoke particles in their greatest numbers, photoelectric detectors respond slightly faster to smoldering fires than ionization detectors.

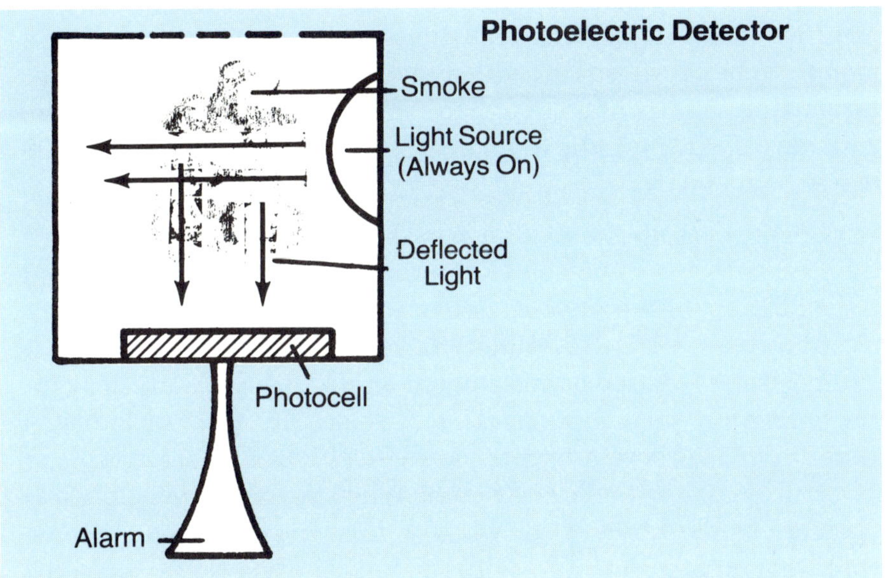

Photoelectric detectors respond to a smoldering fire somewhat more quickly.

Both Types Are Effective

Even though the average particle size changes considerably with temperature, all fires produce a broad range of particle sizes. Therefore, both types of detectors will detect most fires. And while there will be some variation in the detector response time, the differences are fairly small when compared to the amount of escape time the detector provides. Numerous field tests have shown that either type of detector, when correctly installed, will provide adequate warning for escape.

While a comparatively new technology, smoke detectors work extremely well. They have proven effective both in the

When purchasing smoke detectors, only consider those detectors which are listed or approved by a nationally recognized listing organization such as Underwriters' Laboratories, Inc. In addition to UL 217, the testing requirements for UL 167 are recognized as an acceptable and quality standard.

testing laboratory and in those real-life situations where they had to perform. The laboratory test is important for prospective purchasers. Smoke detector buyers should consider only those models listed or approved by a nationally recognized testing organization such as Underwriters' Laboratories, (UL) Inc. The most widely accepted and nationally recognized test requirement for smoke detectors is the UL Standard for Single and Multiple Station Smoke Detectors (UL 217). UL 217 took effect July 5, 1977. (Single station devices sense the fire and then activate a self-contained alarm. Single station devices that can have the alarms connected together are called multiple station.) UL 217 specifies smoke detectors by electronic and mechanical reliability, performance, manufacturing and production, and marking. Instructions must cover operation, installation, maintenance and testing.

UL 217 requires that detectors respond in the sensitivity test to 4 percent smoke obstruction per foot with gray smoke and 10 percent smoke obstruction per foot with black smoke. UL 217 also requires that detectors alarm within specified times to certain room fire tests.

This sensitivity and response time testing is important. In order to pass the UL 217 test, the detectors must be sensitive enough to be of life saving value and must allow smoke to enter the sensing chamber far enough to avoid detection delay. Detectors that pass UL 217 will have an average life expectancy of 15 to 20 years.

Smoke detectors which have passed the UL 217 test standard are good detectors. Look on the testing laboratory label attached to the housing of the smoke detector to see if the device is listed in accordance with UL 217. When this is not apparent, determine if the test standard used is as stringent as UL 217. Secondly, look for the detector's sensitivity setting. A smoke obscuration of 4 percent per foot is required to pass UL 217, but 2 percent smoke obscuration per foot or less is better. When the sensitivity is below 1 percent smoke obscuration per foot, false alarming may become a problem. Smoke detectors with the sensitivity setting between 1 - 2 percent smoke obscuration per foot give good life saving potential with fewer nuisance alarms.

POWER SOURCES

Batteries or household current can power residential smoke detectors. Battery-operated detectors offer the advantage of easy installation; a screwdriver and a few minutes are all that is needed. Battery models are also independent of house power circuits and will operate during power failures. It is critical that only the specific battery recommended by the detector manufacturer be used for replacement. In many instances, a unit with the wrong battery will not respond to smoke even though its test button will function. Additionally, some detectors require special batteries which may be difficult to find and available only by mail order in all but large metropolitan areas. While many detectors on the market take special batteries, more and more detectors require common batteries. If battery-powered detectors are preferred, residents of relatively isolated areas should purchase detectors using batteries which are readily available.

Power failures are relatively infrequent in most urban areas. There is a much higher probability of people becoming complacent and not replacing batteries immediately than of a power outage during the fire either due to utility failure or the fire itself.

Field data indicates that many purchasers of battery-operated detectors have not replaced worn-out batteries immediately. Consequently, most codes requiring detectors in newly constructed homes specify 110-volt, hard-wired units, since people who did not voluntarily purchase detectors are expected to be less knowledgeable about battery replacement. Therefore, the detector powered by household current is usually the most reliable mechanism. However, in some rural areas and areas with high thunderstorm occurrence, power failures may be more frequent, and battery-operated units may be more appropriate.

Several manufacturers produce electrically powered units with stand-by batteries for use during power failures. However, most stand-by battery systems do not last any longer than the primary batteries in battery-operated detectors. While possibly providing peace of mind, these units can be inoperable during power failures if the owners have not replaced a worn-out, stand-by battery. The more expensive battery should last from eight to ten years before needing replacement.

Battery-powered detectors are easily installed. The detector may be attached to the ceiling or wall with adhesives, screws or expansion fasteners, depending on the manufacturer.

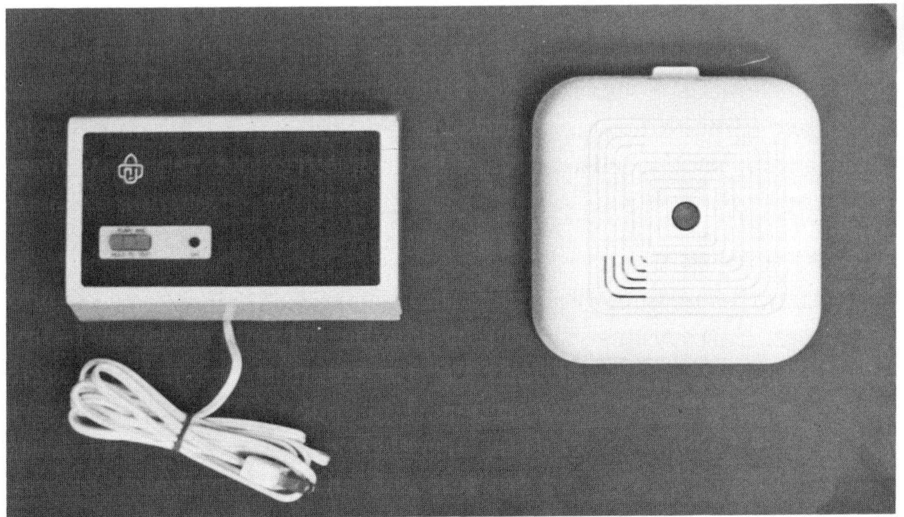

Detectors are powered by either batteries or household current. Both power sources have advantages and disadvantages and the individual purchaser will need to decide which type is the most suitable for his needs.

SELECT LOCATION

A smoke detector in every room will provide the fastest detection times. However, the modest increase in escape time may not be enough to justify the additional expense in some family budgets.

Extensive field tests show that installing a smoke detector on every level of the living unit provides good all-around protection for the least investment. When providing this "every level" detection, the user should consider locations such as bedrooms, hallways, stairways and normal exit routes.

At the minimum, users should install a smoke detector in the hallway outside each sleeping area. The detectors should be close enough to the bedrooms so that the alarm can be heard when the bedroom door is closed. When the bedroom door is closed an additional detector must be inside the bedroom to provide life saving protection in the event of a bedroom fire.

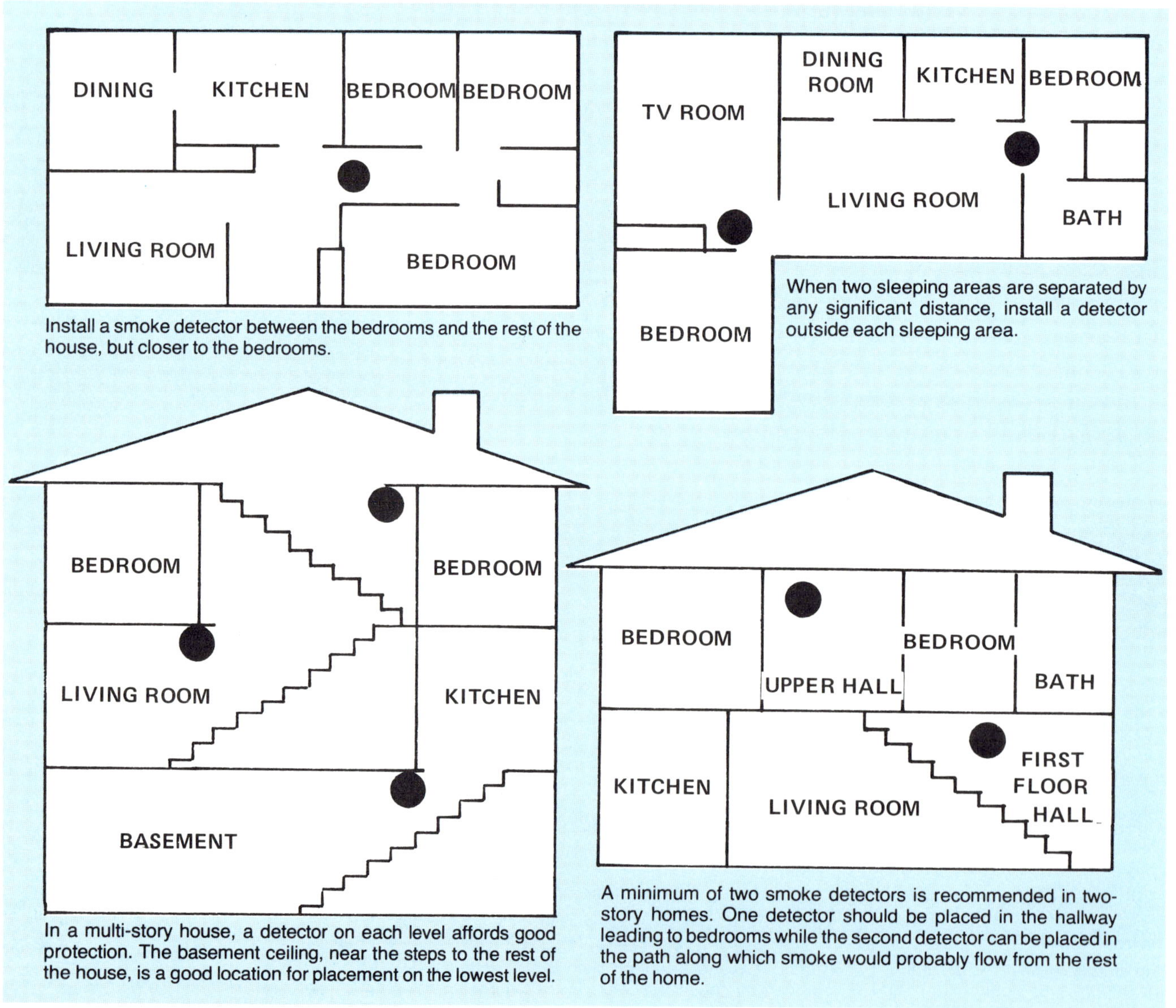

Install a smoke detector between the bedrooms and the rest of the house, but closer to the bedrooms.

When two sleeping areas are separated by any significant distance, install a detector outside each sleeping area.

In a multi-story house, a detector on each level affords good protection. The basement ceiling, near the steps to the rest of the house, is a good location for placement on the lowest level.

A minimum of two smoke detectors is recommended in two-story homes. One detector should be placed in the hallway leading to bedrooms while the second detector can be placed in the path along which smoke would probably flow from the rest of the home.

Correct Placement

Since smoke rises, users should install smoke detectors on ceilings or on walls between six and twelve inches from the ceiling. Avoid placing detectors in the "dead air" high in corners where the wall and ceiling meet.

For proper installation, follow manufacturers' instructions.

Many occupants of dwellings with forced-air systems make the mistake of placing detectors too close to the supply register. As a result, the air stream continuously purges the smoke detector when the ventilation system is on. Avoid these locations by at least three feet. Cold air returns can draw smoke away from the area before it reaches the detector. For this reason, avoid placing detectors between the cold air return and the bedrooms.

The temperature of the mounting surface is critical. For example, the ceiling of an uninsulated attic or of a mobile home can be considerably colder than the room in the winter and much warmer during the summer. This temperature difference can create an air movement barrier preventing smoke from reaching the detector. Therefore, smoke detectors in mobile homes should be placed on an inside wall, never on an outside wall or ceiling. Older and poorly insulated conventional homes may also require detectors to be placed on inside walls or ceilings below heated or insulated spaces.

Radiant panel heating in the ceiling also creates a layer of hot air that may prevent smoke from reaching the detector. Detectors should not be mounted on radiant heated ceilings.

In mobile homes with a standard floor plan, the recommended general location for a smoke detector is on the wall at the end of the corridor entering the common-use areas of the home such as the living room, dining room or family area.

In conventional homes with long central halls, users can increase potential escape time by installing a detector at each end of the hall, or installing a unit approximately every 30 feet.

Once the detector unit is mounted, the battery is placed into the unit and then the unit is tested. Mounting detectors on ceilings takes advantage of the fact that smoke rises.

Wall mounting of detectors should be used when a ceiling is substantially warmer or colder than the rest of the house. Locate the detector between 6 and 12 inches from the ceiling on an inside wall.

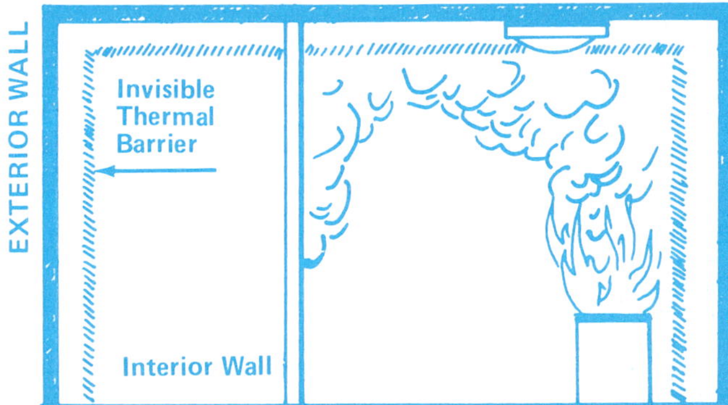

The ceilings in poorly insulated homes can vary in temperature from the remainder of the room. As depicted in this illustration, this temperature difference can create a barrier which prevents smoke from reaching a detector on the ceiling (left illustration). In these cases, the detector should be placed on the wall of an inside room to avoid this barrier (right illustration).

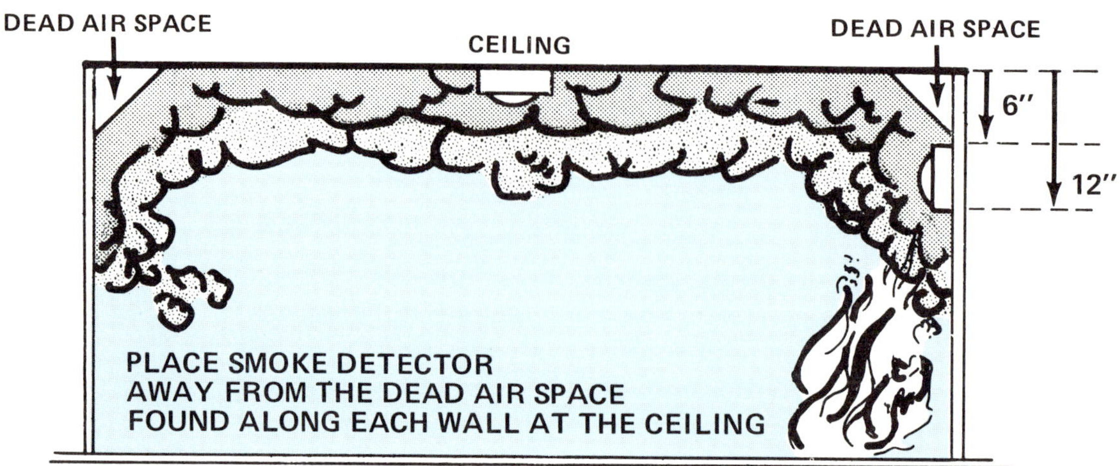

Dead air spaces often exist high in corners where a ceiling and wall meet. Avoid placing detectors in these spaces.

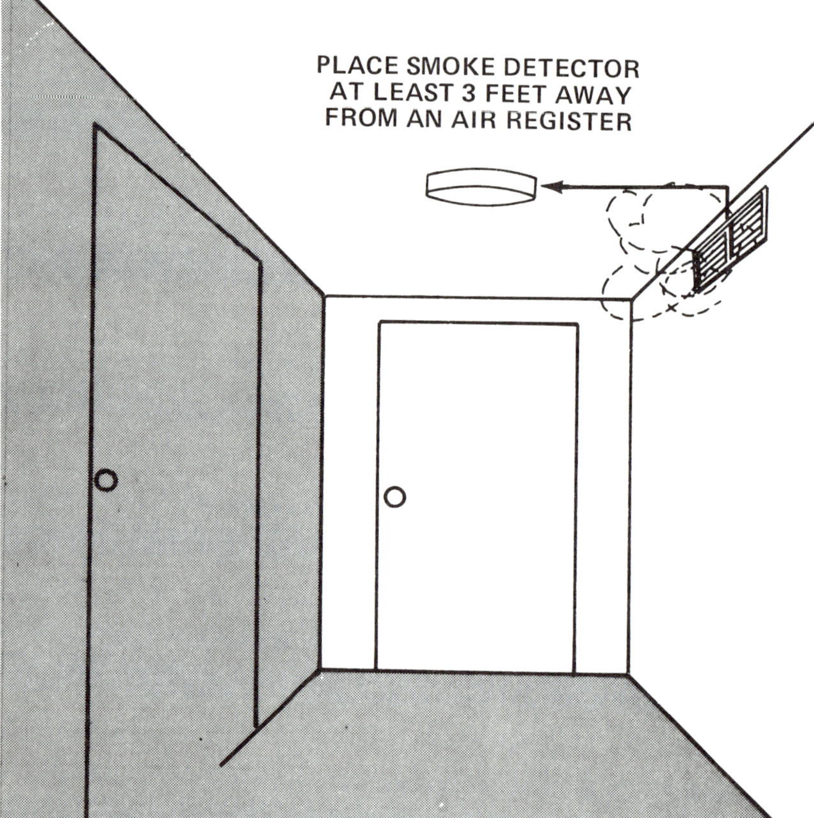

PLACEMENT

Air supply registers can interfere with the effective sensing capacity of a smoke detector. Caution purchasers to locate detector at least three feet away from registers so that the detector will function at its maximum.

Hostile Areas

Smoke detectors are for use in the living and sleeping areas of the home. Areas of the home that are too hot may interfere with electrical circuit operation, while extreme cold can cause batteries to lose power. Dust or grease particles in the air may "confuse" the detector, resulting in false alarms. For maximum service, avoid using smoke detectors in areas that are too hot, too cold, too dusty or too greasy to permit normal smoke detector operation. Examples of hostile areas for smoke detectors may include kitchens, attics, garages, workshops and some furnace rooms. Smoke detectors are sensitive instruments and the manufacturers' recommendations must be followed.

INTERCONNECTING SMOKE DETECTORS

Audibility is critical. If the alarm cannot be heard, a smoke detector does not protect the building occupants. The level of sound necessary to wake a sleeping person varies considerably. The level of background noise, the stage of sleep and physical condition are all important factors. It is questionable whether a detector other than one near the bedrooms will wake sleeping occupants.

To help solve this problem, many detectors can be interconnected so that when one alarms, they all alarm.

The interconnection feature can be found on some household current-powered and battery-powered detectors. Each manufacturer and model must be examined to determine the method used to interconnect the alarms and the capabilities of each device.

Some devices not only connect together but can be connected to an electrically activated switch called a "relay." When the smoke detector senses smoke and turns on all the interconnected alarms, the relay also closes its electrical contacts which can turn on the lights in a deaf person's room, or activate an exterior bell, or whatever switching may be set up. Interconnection is a good feature but it may require a specific request to locate that kind of unit.

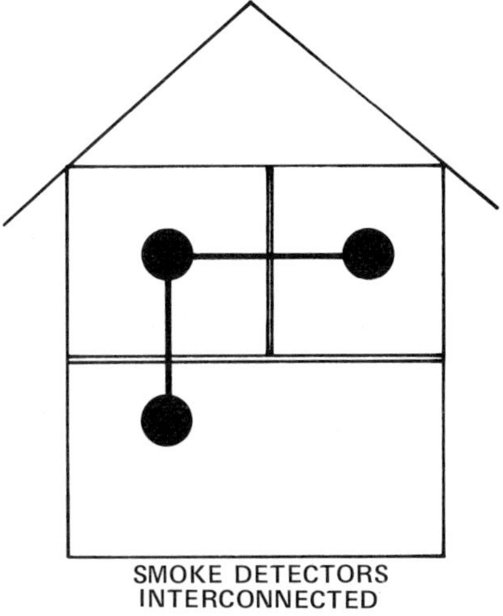

Some smoke detector units can be interconnected so that when one alarms, they all alarm. This interconnection increases the audibility of the alarm.

TESTING

The smoke detector should be tested weekly according to the manufacturer's instructions. It is important that owners test their detectors at least once a month with smoke, because some detectors have test buttons that only check the device's horn circuit. The owner of the detector may not realize this. Only when the detector incorporates a test button that simulates smoke or checks the detector's sensitivity can the "smoke test" be eliminated.

Smoke detectors with test buttons that simulate smoke or check sensitivity should be recommended in homes where "smoke testing" is not likely to be conducted, such as a handicapped or elderly person's home.

Testing with smoke is simple, but suggest safe ways to produce the smoke. Cigarette smoke is most often used for testing. Smoke from burning incense, a small piece of smoldering cotton rope or string in an ashtray can also be used, especially when testing photoelectric detectors. Ionization detectors can also be tested by blowing over the top of the flame on a wood or paper match, directing the invisible or visible smoke particles toward the openings of the smoke detector.

Testing smoke detectors in this manner is only acceptable on newer models with test buttons that simulate smoke or check sensitivity.

Once a month, a detector should be tested. Hold a candle six inches under the detector. To test an ionization detector, let the flame burn. When testing a photoelectric detector, extinguish the candle and let visible smoke drift into the detector. A cigarette can also be effectively used to test a detector.

MAINTENANCE

Owners should conduct maintenance procedures outlined in the owner's manual enclosed with each detector. Cleaning is especially important. Dirt can "confuse" the detector and result in a false alarm or cause the detector to malfunction. Vacuuming and dusting the detector, along with periodic tests, will lessen the chances of false alarm or malfunction.

Major Problem: Batteries and Light Bulbs

The human failure to replace batteries and light bulbs is a major cause of smoke detector malfunction.

The batteries in UL 217-listed smoke detectors will last at least one year when operated according to the manufacturer's instructions. When the battery begins to weaken, the detector gives a warning signal, a chirping sound, at least once a minute for a minimum of seven days.

Several fire deaths have been reported where the smoke detector was in place but not operating because the battery had been removed to stop the battery trouble signal. The fires occurred several weeks later, but the battery had not been replaced.

The light bulb used in some photoelectric smoke detectors is an incandescent lamp. When the filament burns out, the bulb must be replaced. Detectors that use incandescent lamps have at least one spare bulb attached to the detector's housing when packaged by the manufacturer. Light emitting diodes (LED) used in some photoelectric detectors are expected to last the life of the detector.

Selecting a birthday or holiday as the annual date to replace batteries and light bulbs in smoke detectors is suggested.

Follow manufacturer's instructions as to the proper method of cleaning the detector.

Choosing a special date such as a birthday to replace batteries and light bulbs will lessen the likelihood of forgetting this important maintenance task.

Thoroughly cleaning a detector will lessen the chances of false alarms or malfunction.

HEAT DETECTORS

For total residential fire detection, the heat detector should be included with smoke detectors.

For primary life protection, place smoke detectors on every floor level, in each bedroom and in other living areas. Smoke detectors also provide property protection. But smoke detectors do have limits as to operating temperatures and contaminated atmospheres. Examples of problem areas are kitchens, attics, garages, workshops and some furnace rooms. Heat detectors can warn of fires in these areas.

Heat activated switches (the snap-disc buttons or rate-of-rise sensors, for example) can be connected by low voltage wire to some of the multiple station smoke detectors. The result is added protection and a low cost total fire detection and alarm system.

Residential smoke detectors are important but are only the first step in the escape plan crucial to fire safety. The stress situation presented by fire can cause panic and irrational behavior influenced by exposure to low levels of carbon monoxide. A carefully practiced escape plan can reduce panic, counter smoke toxicity and help a family find its way to safety. Each step in planning home escape must involve everyone in the family.

KEY FACTORS FOR ESCAPE

Some key factors in establishing an escape plan include:

- Knowing the alarm sound, and what to do when the alarm sounds in the middle of the night.

- Planning a secondary emergency exit route should the fire be building so fast that the normal exit is cut off. In multi-story homes, this may mean buying an emergency escape ladder to allow escape through a bedroom window.

- Writing down and practicing the escape plan with fire drills in the home. People should act instinctively when the detector sounds an alarm. Everyone can attain this reaction by practicing the fire escape plan with all members of the family.

- Practice crawling to safety to stay under the smoke. Remember before opening doors to touch the door knobs and the top of the door with the back of the hand to test for heat. Use the alternate escape route rather than opening a hot door.

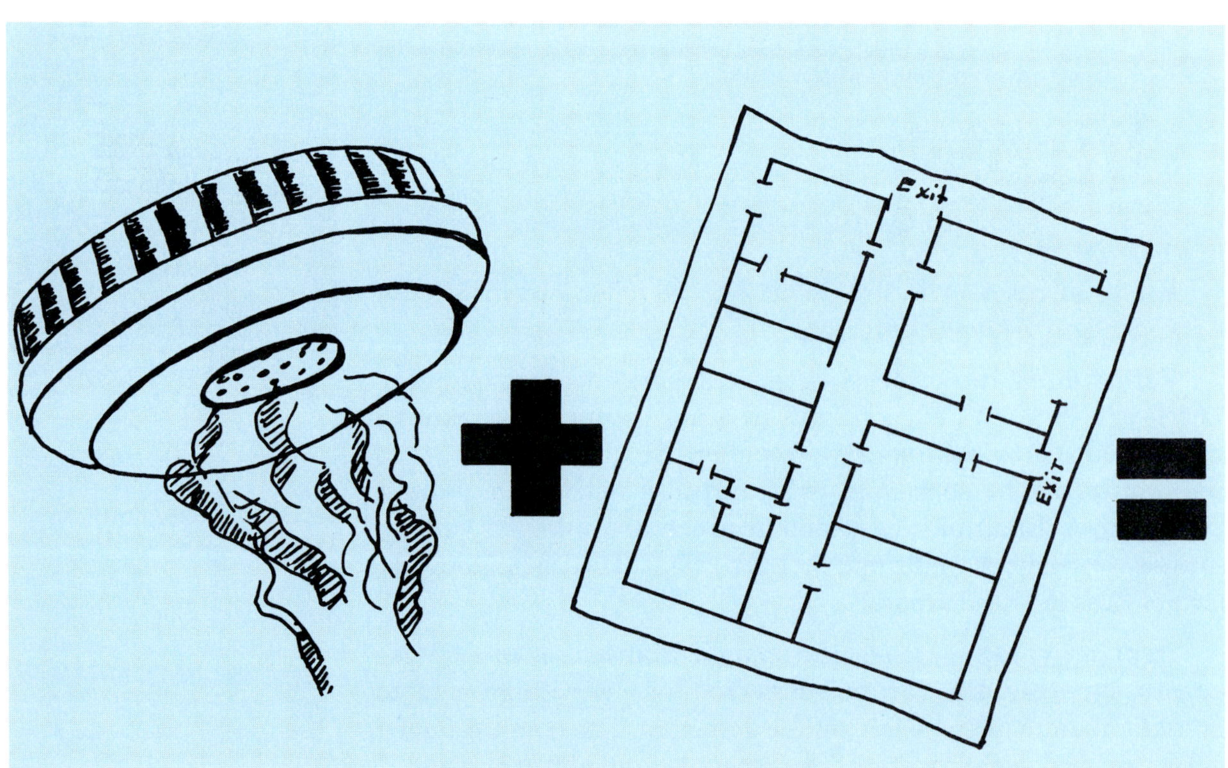

- Immediately leave and go to a prearranged meeting place outside the house when the detector alarms. A tree, street light or a neighbor's home can be the meeting place. This is very important since numerous people have been killed when they have gone back into a burning house looking for someone who has already escaped.

- Never stop to call a fire department from inside the home. Fires can build at amazing speed and the occupant may be cut off or even killed while on the phone. Call the fire department from a neighbor's home.

- Never re-enter the home once you have left.

ACT NOW

It is necessary to educate the public about smoke detectors. Use the five step planning process for public fire education. Get involved with resource exchange of smoke detector programs and materials wherever a successful program exists.

The reward is knowing that the potential of saving a life or preventing a fire injury is very real. Fire fighting through public education doesn't get many hero awards, but when someone explains how the information provided helped protect them from their fire, you will silently know that it has been worth it all.

SUGGESTED LESSON PLAN FOR FIRE EDUCATION PROGRAM

Background

Let us first look at some significant facts on the value of smoke detectors. Over 90 percent of the nation's building fire deaths occur in residential buildings, according to estimates by the National Fire Data Center of the National Fire Administration.

Household fires are especially dangerous at night when the occupants are asleep. Most casualties are victims of smoke and toxic gases.

The National Fire Administration has encouraged the installation of smoke detectors in all of the nation's homes. Smoke detectors have the potential of reducing home fire deaths by over 40 percent.

Smoke detectors help the home occupants by giving early warning so they can escape. Also, the earlier the fire is discovered, the less property the fire can destroy before it is extinguished. In two very important ways, the smoke detector can also help the fire department.

- The alarm can warn of fire while it is still small, making extinguishment easier.
- When the dwelling occupants escape from a burning home because of an early warning, the firefighter does not have to attempt an unnecessary rescue.

Daily, the fire services, community groups and the smoke detector industry are bringing all these facts to the public's attention.

The result is that many communities have adopted legislation requiring smoke detectors in all new and existing homes, increasing the need for smoke detector information.

Instructor's Notes

Display (Note: this includes houses, apartments, hotels, mobile homes, etc. Only about 6 percent occur in schools, hospitals, nursing homes, factories, etc.)

Cite local examples illustrating these points.

Display: Smoke detectors can reduce the nation's home fire deaths by 40 percent.

Examples: Fire department pamphlets, newspaper articles, advertisements.

Refer to an example appropriate to your area.

Ranging in price from $15 to $50, smoke detectors are reaching all time high sales. It is estimated that 10 to 12 million detectors will be sold in the United States in 1978 (compared to 50,000 in 1972).

Clearly, people are interested in smoke detectors, and there are two questions they most often ask:

- Which type of detector is best?
- Where should the detectors be located in the home?

Types Of Detectors

When selecting a smoke detector, your first consideration is listing by a nationally recognized testing laboratory, such as Underwriters' Laboratories, Inc.

Types Of Detectors

There are two types of household smoke detectors: photoelectric and ionization.

Photoelectric Detectors

Let's first examine the photoelectric smoke detector. It uses a light source and light sensitive cell in a darkened chamber. The light source produces a light beam. The photoelectric detector goes into alarm when smoke enters the darkened chamber and reflects the light beam into the light sensitive cell.

Ionization Detectors

The other type of home smoke detector is the ionization detector. The ionization smoke detector uses a radioactive material to make the air within the sensing chamber conduct electricity. The alarm sounds when smoke interferes with the flow of the electrical current in the ionization chamber.

Display

Display: Sample of the UL label. Show where it is found on your demonstration smoke detector.

Display or distribute copies of schematic rendering of a photoelectric detector. Explain: Cigarette smoke visible in the light from a projector illustrates how smoke reflects or scatters light. Explain: The light source is either a small incandescent light bulb or an LED (light emitting diode) similar to those used in electronic watches.

Display or distribute copies of schematic rendering of ionization detector.

The radiation source presents no health hazard to the home's occupants.

Explain: The U.S. Nuclear Regulatory Commission states that a person flying round trip across the United States receives about five millirems of radiation because of the increased dosage at high altitudes. This exposure is ten times greater than the exposure you would receive by holding the detector next to you eight hours a day for an entire year.

Which is best? Ionization detectors best sense the fires that produce very small or invisible smoke particles. Photoelectric detectors sense smoky smoldering fires best — those that produce larger smoke particles. But let us test them.

As you can observe, detectors alarm quickly. Since most home fires produce both visible and invisible smoke particles, the National Bureau of Standards has concluded that properly installed photoelectric or ionization detectors are both adequate in home fire life saving potential.

Do a simple smoke test with each type of detector, if possible.

Battery Units

Battery-operated units give round-the-clock protection, if a responsible person has maintained the battery. All UL-listed battery units are designed so the batteries last at least one year, if operated according to manufacturer's instructions. As the battery weakens and needs replacement, a warning chirp or beeping sound occurs at least once a minute for a minimum of seven days. Some units have a red flag to signal a weak battery.

Display: Battery-operated type detector is available. Show the battery in demonstration unit.

If possible demonstrate with a weak battery. Battery-operated units can be either photoelectric or ionization.

Electric Plug-In Units

Some electric powered units are equipped with a cord which plugs into an electric outlet. Be sure the outlet used cannot be turned off by a wall switch and is not on the same circuit as heavy electrical appliances that may blow a fuse if they catch fire.

Display: Electric plug-in type detector if available.

Electric Wired-In Units

Some electric units may be directly wired into the home electrical system. This will usually require the services of an electrician.

Display: Electric wire-in type detector if available.

AC-DC Units

Electrical units with battery backup power are available. As long as the battery is maintained, this secondary power supply keeps the detector operating if the electric power fails.

Display: Electric unit with battery backup if available.

Which is Best?

To answer the question, "Which smoke detector is best?", we can say: "If a photoelectric or ionization detector is listed by a nationally recognized testing laboratory, it will be an effective home life safety device, *If* it is *properly installed and maintained."*

Explain: There is not a "good, better or best" rating for detectors listed by a nationally recognized testing laboratory. Smoke detectors either pass or fail the laboratory test standards.

Placement

The most important location to install the smoke detector is the hallway outside the sleeping area. Homes with more than one group of bedrooms should have detectors installed at each sleeping area.

Display or distribute copies of every-level protection diagram.

Homes with more than one floor level should have a smoke detector on every level. Installing a smoke detector in each bedroom increases protection particularly when there are smokers in the home. Here are some facts that illustrate the life saving protection of smoke detectors.

Display or distribute Life Safety Index.

Data from tests conducted for the National Bureau of Standards has been used to develop a "life safety index." A single smoke detector in the sleeping area provided over three minutes to escape through the normal exit in over 35 percent of the test fires. When smoke detectors were on every floor level, an alarm allowed the desired three minutes for escape time in an impressive 89 percent of the test fires. That is significant life saving protection at a reasonable cost.

Note: This is potential warning to allow escape. To safely escape the alarm must be heard (may require interconnecting remote detectors to bedroom area alarm) and occupants must react properly.

Interconnecting Units

Detectors remote from the bedrooms may be difficult to hear, so smoke detectors that can be interconnected are suggested. Then all detectors sound the alarm when one detector senses smoke.

The number of detectors needed will depend on the residence. If a family can only afford one smoke detector for their house or apartment, it should be located in the hallway outside the bedrooms.

Smoke Detector Placement

When installing a smoke detector on the ceiling, it should be placed near the center of the hall or room. When placing a smoke detector on the wall, locate it six to twelve inches from the ceiling.

Avoid the *dead air* space at the top edge of the room — where the ceiling and wall meet — because air flow patterns can prevent smoke particles from reaching the corners in the early stages of the fire.

Exterior Walls and Ceilings

Do not install a smoke detector on a poorly insulated exterior wall or ceiling, such as are often found in mobile homes or light wood frame construction. Extreme exterior temperatures can cause a thermal barrier on the inside which prevents smoke from reaching detectors mounted on the poorly insulated surface. If poor installation is suspected, install the detector on an interior wall six to twelve inches from the ceiling.

Air Supply Registers

Avoid placing detectors within three feet of an air supply register or return. Smoke could be pushed or pulled away from the detector by the air flow.

Display or distribute copies of interconnected diagram.

Explain: Interconnected units may require installation by an electrician.

Display or distribute copies of placement diagram. Note: Individual detectors with highly directional smoke entry openings may function differently when installed on the ceiling or wall. Follow the manufacturer's directions.

Display or distribute copies of dead air space diagram.

Display or distribute copies of poorly insulated surface with smoke diagram.

Display or distribute copies of poorly insulated surface placement diagram.

Display or distribute copies of air register supply replacement diagram.

False Alarms

In order to minimize false alarms, avoid installing the smoke detector where it will be exposed to fumes from cooking or furnace, or fireplace smoke or dust.

Maintenance

Most smoke detectors require very little maintenance, yet that maintenance is important since the human failure to replace batteries and bulbs is a major cause for detector malfunction.

Explain: A spare light bulb comes with the detectors that use them. Always replace the spare bulb immediately when it is used.

Replace batteries yearly or whenever the weak battery signal sounds.

Some photoelectric detectors use an incandescent light bulb which will need replacement. Replace the light bulb when the warning signal indicates it has burned out.

Vacuum or dust the detector periodically. This will lessen the chance of false alarm or malfunction.

Testing

Some detectors have a test button that will activate the alarm, while other units have a power-on signal only. All smoke detectors should be tested at least once a month with *smoke*. Detectors should also be tested weekly in accordance with manufacturer's instructions.

Test demonstration detectors. Explain: Detectors are electronic devices like radios and TVs that can wear out and fail. Testing is the only way to be continually certain this has not happened.

Always test the smoke detector when returning from vacation or when noone has been in the home for several days. Explain: Battery trouble signal may have sounded while the family was away.

DETECTORS NOT ENOUGH

We have talked about the smoke detector as a life saving device. But you must remember that a smoke detector is *only* a device, a machine. A smoke detector cannot put out a fire. It cannot get people *out* of a fire.

Just having a smoke detector properly installed and maintained is not enough. You must know what to do when a fire occurs in your home. The family must have an *Escape Plan*.

Display: Smoke detectors are not enough.

The Escape Plan

All family members should know *two exits from every room*, so safe evacuation can be quickly accomplished even when the normal exit path is blocked.

Display: Family escape plan.

Display: Two exits from every room.
1. Door
2. Window

Everyone should then assemble at a pre-determined location outside, call the fire department from a neighbor's phone and let the fire department fight the fire.

Display: Assemble outside at pre-determined location.

Display: Call the fire department using a neighbor's phone.

Practice the escape plan so that in an emergency the proper response (evacuation) will come naturally.

Display: Practice the escape plan.

Escape planning must be part of your smoke detector protection. The alternative is a false sense of security which can lead to tragedy.

Explain: Home fire drills should be initiated by sounding the smoke detector's alarm. Fire drills may sound silly, but they result in saving lives.

REVIEW

Can you remember the two questions about smoke detectors that are most often to be asked?

Have the class respond.
1. *Which type of smoke detector is best?*
2. *Where should the smoke detector be located?*

Can you remember the answers?

Have the class respond.
1. *Photoelectric and ionization type detectors are both adequate in life saving potential when they have been listed by a recognized testing laboratory.*
2. *Must provide a smoke detector in each bedroom hallway and there should be at least one smoke detector on each floor level of the home.*

Display

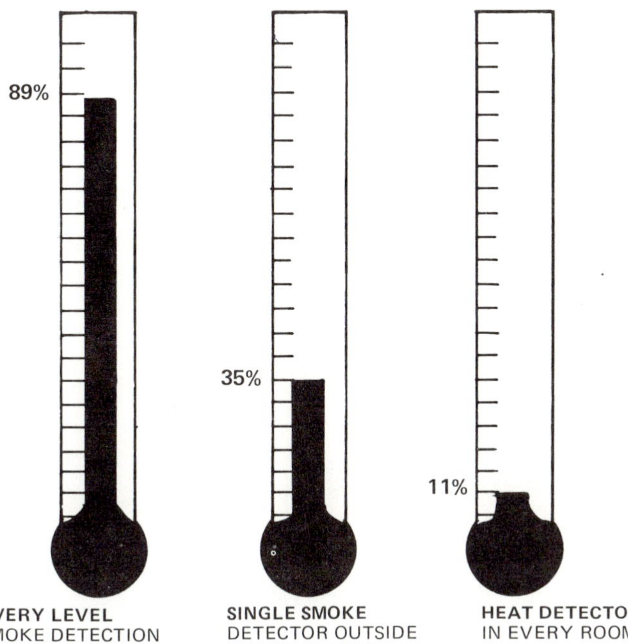

The life safety index for three-minute escape time shows the value of installing smoke detectors on every level. Notice that a single smoke detector outside a sleeping area is three times as effective as a heat detector in every room.

Smoke detectors come in a variety of designs to complement room decor. Only purchase detectors that are listed by a nationally recognized testing laboratory. Underwriters' Laboratories and Factory Mutual, Inc. are two such recognized testing facilities.

SUMMARY STATEMENT

1. Buy a listed detector.

2. Install it correctly.

3. Test and maintain regularly.

4. Make and practice an escape plan.

They can save lives. Perhaps even yours.

It is important to educate firefighters about smoke detectors. Many times the public will call the fire department when they have questions as to which type of detector is best or where to locate detectors in their homes.

5

Working With The Media

Among the most beneficial of community resources readily available to help the fire service get its message to the public is the popular media: TV, radio, newspapers and magazines. In the planning process, inventory the available media in the community and surrounding areas. Statewide radio news networks, television stations with broad coverage, state and local newspapers and even billboards should be considered as possible sources of contact with the public.

Even though all these means of providing this contact are collectively known as "the media," each one has a slightly different function and audience, and each has its own method of receiving news and public interest messages and methods of transfer. Differences are also evident in the internal operations of each organization.

Generally speaking, each news medium is obligated to some extent to provide public service messages to its audience at a no cost basis to the source or supplier of the information. It is through this obligation that the media becomes a valuable ally to the fire safety education program. Remember, however, that the relationship must be a mutually cooperative one in which the fire service, the media and the public benefit.

The first consideration in forming a sound relationship with the media is "getting to know them." Seek out editors, reporters, news team anchor personnel, and those others involved with news story or public safety announcement "copy." Express sincere interest in their operations and services so you will better understand how they may help you. Get the ideas of what you want them to do for you clear in your own mind first. In other words, identify your objectives so the media personnel can fully understand your request and the proposed benefits of the message you want them to present.

Invite suggestions from them as to the type of material they will accept, the manner in which the material should be prepared and how far ahead the material should be submitted in order to assure a timely release date. It may also be advisable to learn the proper routing of "copy" through the various departments in a news service and how priorities are assigned to the PSA's (public service announcements).

The media is an invaluable asset to the fire prevention educator. Messages about fire prevention can be spread to a large number of people at little or no expense. It is important that a good rapport be established with media personnel so they are receptive and cooperative when working with the fire department.

MATCH THE MEDIA AND THE MESSAGE

Another consideration when seeking news or press coverage is to choose the media that will be most beneficial to the message you wish to present. It would not be productive to use a newspaper or a television news broadcast to direct a message about fire safety to children. Likewise, a target audience of the elderly would receive little information concerning heating safety from a pamphlet distribution campaign at a shopping center. In the process of designing the specific messages, remember to consider what media resources are more likely to be used by the chosen target audience.

A spirit of cooperation must be generated and nurtured between the department and the media. In the "give and take" relationship between two organizations, each tries to maintain the integrity of its structure and purpose while helping the other. Even though you realize the importance of the message, convince the media of its benefits and assure them that the message will be presented in a quality way. Make sure that the message or story you want released to the public will be of mutual benefit to both parties involved. Show a sincere concern for the reputation of the newspaper, television or radio station and avoid asking them to run a story that would later embarrass them or be detrimental to their purposes.

Radio is another means of communicating with the public. It is up to the fire prevention educator to initiate contact and to gain the support of local radio stations.

Local television stations may carry special programs on fire prevention and public education upon request. Have a program thoroughly coordinated and choose a person who can communicate well to deliver the program. Without a qualified speaker to deliver the program, the message may be ineffective.

- Be certain that the material is news or of importance to the public. (Some of the things that *may* make news for a public servant include: speeches, articles written, appointment of committees, election of officers, action taken at meetings, visiting officials, important interviews, job upgradings, service statistics and reports, and safety-oriented reminders for the public.)

- Notify media well in advance when presentations, forums and addresses are to be held, so that they can arrange to cover them and plan for space. Invite the media to important luncheons, dinners and meetings. Timely feature stories, especially in Sunday newspaper editions, should be sought. In most instances, Sunday and Monday morning release dates often result in adequate and prominent space being given to stories. When possible, such copy should reach the editors not later than Friday but establish particular guidelines from your own local contacts.

- Special stories, deserving special attention, should be outlined to the editors with inquiries as to whether the full facts may be given to reporters.

General Tips Fo

- Get *fresh* stories to the papers.

- See reporters promptly when they call. Show no favoritism. Give identical stories to all of the papers at the same time, with the same release date.

- Do not editorialize or give opinions in straight news releases. Avoid making "off-the-record" statements (joking or uttering facetious remarks) to press representatives. These can lessen or completely destroy your relationship with them. Avoid unsupported claims on controversial subjects.

- When press photographers attend meetings, cooperate with them in assembling groups for pictures. Do not insist on giving your opinion as to what makes a good picture.

- Learn if there are opportunities for pictures to be taken by the media photographer or taken by the department's photographer and given to the media. If possible, when important events occur, have action-camera shots taken.

- Always have an objective for your publicity. When press space is "tight," give the newspapers the most important and beneficial angle.

- Names, job titles, addresses and other information must be correct and complete.

- Never "fight" with the media or point out their mistakes to force publicity.

- Be sure all releases are cleared with your superior or person responsible for press articles.

- Be polite and express thanks for adequate press coverage and treatment.

If the fire department is to write and release its own copy:

- Outline the news story before writing it.

- Releases must answer the questions Who, What, When, Where and How.

- Do not attempt to create headlines. Wording is almost always done later by editors who must consider the actual space allowed for the headline.

Working with the Media

- Type news stories on paper 8½ inches wide by 11 inches long.

- Double-space or triple-space between lines.

- Start the first page half-way down the sheet. Editors need the top half of the primary page for instructions to the copy desk and typesetter.

- On page 1, upper right-hand corner, give the source of the story, including name, title, department address and telephone number.

- In the upper left-hand corner, put release date, such as : For release: In newspapers on Friday, October 8. Or: For release on or after Friday, October 8. Or: For immediate release.

- At the top of each sheet, place a "slug" - two or three words that will indicate to the editor what the release is about with a quick glance and will help in case the sheets become separated on his desk. For example, "Local Fire Inspector Attends IFSTA Validation Conference."

- Do *not* submit carbon copies to any of the newspapers. Retype instead for the personal effect.

- Number each page and indicate on the same line as the number what the story is about. For example, -4- "Local Fire Inspector Attends IFSTA Validation Conference."

- Use only one side of the paper.

- Have one inch margins.

- Place an "end" mark at the close of copy, such as -30- or #. These are journalist's codes which will add to the professional look of the release.

- Present neat copy without "strike-overs" or "eraser-blurs."

- Give full and accurately spelled names of people. For example, "Those assisting were Mr. Jake R. Wilson, Mrs. Frank S. Stell, and Miss Nellie Bly." Use initials if they are given to you.

When writing copy for newspapers, follow these guidelines

- Make the "lead" (the first) paragraph serve as a thumbnail sketch of the story. Write the story so that, if necessary, it can be cut to fit the space limitation which may exist. Have the less important *details* in the last few paragraphs so that copy contraction will not wreck the continuity or completeness of the story.

- Let the expected publication date govern the use of "yesterday," "today" and "tomorrow."

- If extensive quotes of a speech are to be incorporated in an article, they must be exact. Obtain the written speech prior to writing the story if actual wording is unclear. When statements are made be sure to credit them to the right person.

- Do not load the story with statistics.

- Briefly, interpret your statistics.

- Furnish press photographers with correct names of the people in each picture in the order in which they are standing or sitting. For example: "Top row, left to right: Mrs. Carlton J. Beale ...," "Second row, left to right: Mr. William K. Lyons..."

- When submitting a photograph for publication, attach a typed caption. Do not write on the back with a hard pen or pencil.

- Employ short words, short sentences and short paragraphs.

- Write as if speaking. Present your facts in language which can be understood.

ACKNOWLEDGE ASSISTANCE

Supplying the media with information concerning the effectiveness of your programs will help strengthen the department's relationship with the media. So often, newspapers or television and radio stations are willing to help disseminate information to the public yet never receive thanks or follow-up information concerning the project.

Remember: Once the communication and friendship have been established with various media, the department must periodically reinforce the relations in a spirit of cooperation rather than out of a necessity of getting the fire department's message to the public. In other words, treat the media as friends for they can truly be a strong bridge between the fire department and those it serves.

Evaluation is an essential element of a complete program. How well did you do? How was your program received? How many people were reached? Has your program done what it was designed to do? These are questions that you ask of your program and this information could also be supplied to media personnel who may be willing to help in the evaluation process or counsel your department on the strengths and weaknesses of the program.

SAMPLES OF MEDIA MESSAGES

Enclosed are several pages of public service announcements concerning fire prevention. Fire prevention is our concern not just one week per year but all year long.

Please use these public service announcements as much as possible during FIRE PREVENTION WEEK and all through the year.

Thank you for your cooperation.

RADIO SPOT ANNOUNCEMENTS

To help make your community safer . . . for use during FIRE PREVENTION WEEK . . . and throughout the year.

(30-Second Announcement — 64 words)

Have your kids started to plan their Halloween costumes? Often those costumes are made of the flimsiest materials . . . so be sure they are flame-proofed before the youngsters put them on. A simple way is to dip them in a solution of 4 ounces of boric acid and 9 ounces of borax to a gallon of water. Don't let a flame or a spark turn your children's Halloween into a time for tragedy!

****

(50-Second Announcement — 100 words)

Cold weather is coming, and a fire's going to feel good before long! But it wouldn't feel good if that fire were burning down your home . . . and if you don't take precautions, that can happen! So have your furnace, chimneys and flues inspected and cleaned if necessary before the time comes to use them. See that any heaters you own are clean and in good repair before you turn them on. Provide a screen for your fireplace, and see that it is always in place whenever you light an open fire. Make sure that all the fires that burn in your home are *friendly* fires!

(1-Minute Announcement — 122 words)

Leaves are already starting to fall . . . which means it's about time to get out the rake and go to work on them. After you've got them raked up, you'll be burning them if your town allows it. And for you leaf-burners, here are a few words of caution. First, wait for a windless day before you light the fire. Second, connect up your garden hose and have it handy. Third, put the leaves in a wire mesh container with a cover. Fourth, set it up well away from buildings or fences. Fifth, keep children away from it. And sixth, stay with your fire until it is not only out, but cool to the touch. Fire out of control is a terrifying, destructive thing. Keep your home and family safe.

SHORT PSA'S FOR RADIO OR FILLERS FOR NEWSPAPERS

1. Another Fire Prevention tip from your Fire Department. Learn not to burn! Get rid of extension cords and multiplying outlets.

2. Put smoke detectors near the bedroom says the . . . . Fire Training Academy. Give the kids *time* to get out.

3. Another Fire Prevention tip from the . . . . Fire Training Academy. Plan two ways out of each room. Have a meeting place. Have a fire drill.

4. Can you find fire exits in the dark? Do you know where the nearest outside phone is? Find out. Learn not to burn.

5. Watch the weather and wind. Don't let trash fires get out.

6. Keep the fire in the fireplace. Have your screen large enough for the hearth.

7. Store matches in metal containers. Keep them away from children and heat. Fires don't just start — they're started.

8. Spring clean-up is fire safety time. Have a fire safety check-up and clean-up.

9. Gasoline explodes. If you can smell it, it's dangerous. Keep it ventilated.

10. Keep your children fire safe. Learn not to let them burn. Call your Fire Department or the . . . . Fire Training Academy.

(20-Second Announcements)

Like an open fire on a winter evening? Fine . . . but remember, sparks from that fire can start another fire — on the roof! Be sure your chimney is free of cracks and the roof shingles are in good condition. Keep your open fires where they belong: in the fireplace!

* * * * * * * * * * * * * * * *

Fire is the 2nd leading cause of accidental death in Arkansas. Most of these fires, 82 percent, occur in homes. Almost half of the victims are under 10 years old or over 65 years old. Check for fire hazards in your house. Look in the attic, the basement, closets, all storage areas. The Fire Training Academy urges you to please be careful with fire in your home.

* * * * * * * * * * * * * * * *

A child wakes up in the night. There is smoke in the room . . . the house is on fire! Will the child know what to do? Does each person at your house know what to do in case of fire? Plan ahead. If fire does start, be sure each person knows how to get out. This message a courtesy of the (   ) and your local fire department.

PUBLIC SERVICE ANNOUNCEMENTS

SUBJECT: Fire Prevention
FROM: Your Fire Department
PROGRAM: Christmas Holiday Season

(20-Second Announcement)

Decided what to give the children this Christmas? Whatever toys you select, make sure they're *safe*. Look for the UL label of fire and shock safety on all electrical and heat-producing toys. Choose chemistry sets carefully. Supervise play with toys involving fuels and chemicals. Protect your youngsters from Fire!

****

(30-Second Announcement)

Is anyone in your family taking part in a Christmas pageant or choir recital during this holiday season? Costumes and choir robes are often made of loosely woven material which is highly combustible. No one wearing such a garment should ever get close to a flame or source of heat. If candles are to be carried or used as decorations, they should be electric candles. Don't let Fire turn Christmas joy into tragedy!

****

(1-Minute Announcement)

Are you planning a holiday party in your home, office, church or school? Holiday parties often center around a Christmas tree — and the Arkansas Fire Training Academy reminds you that evergreen trees can catch fire easily and burn furiously. So make safety as much a part of your party plans as the decorations or refreshments. Take these precautions. First, don't set the tree near a stairway or elevator shaft which would produce a draft. Don't let it block a door or any exit. Provide plenty of ashtrays for the smokers — and don't allow smoking near the tree. See that all decorations in the room have been flame-proofed. Have someone in authority inspect the tree every day to determine whether it should be left up any longer. Don't let Fire come to your party!

(1-Minute Announcement)

Christmas is coming, bringing with it a serious home fire hazard: the Christmas tree. Evergreen trees ignite easily and burn rapidly . . . so for safety's sake, choose a small tree and keep it outdoors until just before Christmas. When you bring it in keep it moist and cool, away from any possible source of heat or sparks, such as heaters or fireplaces, electric trains or light switches. Use only fireproof decorations. Be sure all lighting sets and electric cords are in good condition and bear the UL label of fire and shock safety. While your tree is up, check periodically to see if needles near the lights are turning brown. If so, move the lights. And when needles start to drop, take your tree down and discard it — outdoors. Keep your Christmas merry!

(5-Second Announcements)

"Learn Not To Burn" says . . . . Fire Training Academy. Have your heaters and furnaces professionally checked each fall or winter.

****

Learn Not To Burn. Have a fire drill tonight! At home or work keep folks alert.

****

Store grease away from the stove. Clean up spilled or collected grease. Prevent fires.

****

(10-Second Announcement)

The time is 5:05, time for a safety reminder. Open fires are a serious winter hazard. Always keep a sturdy fire screen in place. Stop fires!

****

(15-Second Announcement)

(Fifty Arkansans) have died in the first seven months of this year — not from cancer — not from automobile accidents — from *fire*. Fire is (Arkansas') second leading cause of accidental death. The XYZ Company and your local fire department urge you to "Learn Not To Burn."

****

PUBLIC SERVICE ANNOUNCEMENTS

SUBJECT: Fire Prevention
FROM: Your Fire Department
PROGRAM: Sports

A good athlete always checks his equipment before a game. YOU should check your home for fire hazards — before you have a fire. Here are a few things you should check right now . . . before fire strikes:

* Do you have enough ashtrays?

* Have you only 15 amp fuses in your electric circuits?

* Have you cleaned out the rubbish or litter in your attic or cellar?

* Check your home for fire hazards now — before you have a fire.

****

Football season is in the fall. Fire season is the year around. Within the next ten minutes, at least ten American families will be threatened by fire. A *third* of these fires will be caused by cigarettes and matches.

* Does your home have enough ashtrays?

* Do you or any of your family smoke in bed?

* Practice fire safety with matches and cigarettes and live a little longer.

****

We have all heard the expression — win or lose, be a good sport. When you suffer from a fire in your home — you always *lose* because fire is *never* a good sport. Check your home carefully before taking a ride or retiring for the night. Look for burning cigarettes. Remove all rubbish or litter that might catch on fire. Finish your last cigarette *before* you go to bed.

Don't be a good sport with fire. Stamp out all the fire hazards in your home.

Smoke kills.
Home smoke detectors save lives. Learn to stay alive.

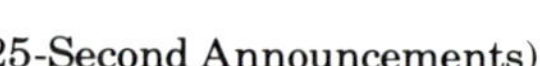

(25-Second Announcements)

Matches start fires. Right? Wrong! People start fires and they have many ways of doing it. Quite often people don't even need a match. In Arkansas each year many homes are burned and people lose their lives because of carelessness. Be careful with fire at your house. This message brought to you by the (   ) and your local fire department.

****

A stove, a water heater, a cigarette or a bad electrical connection could start a fire that would destroy your home. Paper and cloth are two materials that burn quickly. A fire may start in the upholstery of a chair, a child's nightgown or closet. Please, be careful with fire at your house. This announcement is a service of the (   ) and your local fire department.

(30-Second Announcement)

Cold weather's here — and as the temperature drops, fire losses climb! Do you know what to do if fire strikes? First, get everybody out of the house. Second, call the fire department. Plan now for a possible future emergency. Figure out two routes to the ground from every upstairs bedroom . . . and teach every member of your family how to contact the fire department. Then, if fire strikes, you can strike back!

(1-Minute Announcement)

Do you have a child between the ages of 1 and 14? Do you know that every year accidents take the lives of over 11,000 children in that age group? That's over three times the child death rate for any illness! In the home, the top child accident killer is fire . . . and since December and January are the worst months for fire deaths, the . . . . Fire Training Academy urges you to begin today taking the following safety precautions. First, make your home as fire safe as possible . . . and keep hazards like matches, hot liquids and electric appliances out of youngsters' reach. Second, make safety a part of each child's daily training. Remember, fire hazards may be an old story to you, but they're a new story to every child. Give that story a happy ending!

Using Visuals

Throughout this manual, the use of visuals in public education programs has been stressed. Even the most important educational message must compete for the attention of a busy, distracted public. For this reason, fire department educators have augmented their presentations with visual aids of the hazards their message addresses. Besides holding the attention of the public, the use of visuals fulfills an even greater role. Research has pointed out that approximately 80 percent of all that is learned is learned through the sense of sight. This is an important point to remember when preparing a public fire education program. A lecture or speech will convey the message, but when the talk is supplemented by visuals, its effectiveness is increased and the message will probably be more vividly imprinted.

Visuals do not necessarily need to be elaborate or expensive. Public fire education material incorporating visuals, and some audio, is now available at reasonable costs. If an existing program does not meet particular requirements, a presentation using visuals can almost always be prepared without a great deal of expense. This chapter will cover in detail the use of visuals in slide programs and in table top demonstrations. Other visuals can be used, depending on the fire department's needs and finances. Simple posters, cut-out felt flames for children to use when practicing stop, drop and roll, and other such visuals are effective and their cost is minimal. Other visuals such as films contain much good information and can very often be secured for a nominal fee. Consider all visuals that are available or have the potential of successfully supplementing the program and choose those that best fit local needs.

SLIDE PROGRAMS

The first requisite in preparing a slide program is to select a goal or to decide on what is to be accomplished by the program. Define the broad purpose of the program. Some goals of a slide program might include the following:

- The viewer will have a better understanding of the consequences of inadequate fire safety precautions — monetary loss, physical injury or death.

- The viewer will better understand the immense need for fire prevention.

- The viewer will better understand what constitutes a fire hazard.

- The viewer will learn what his/her action would be in the event of fire.

The goal selected could be to educate the public on the advantages of purchasing home smoke detectors or to educate the public on home fire hazards.

DETERMINE GOALS, OBJECTIVES AND SCOPE OF PROGRAM

Once the broad goal has been established, define more restricted objectives. These objectives should result from studying fire problems in the area. If it has been determined that careless smoking habits and flammable liquids contribute to many home fires, this area may be chosen as the core of the slide program. If the goal is to alert the public to the advantages of home smoke detectors, objectives could include showing proper maintenance and testing, pointing out the differences between ionization and photoelectric detectors and so forth. Once objectives for the slide program have been decided on, examine the scope of the program. Will general problems and needs, or a special, limited problem be addressed? One program may be to present a total package on smoke detectors including the success of escape made possible by early warning devices. A more specific program may be approached by concentrating on smoke detectors in mobile homes. This program could go into placement details and the maintenance and testing requirements.

The next step in creating a slide program is determining the audience for whom the program is intended. Will the program be shown to children, adults or special groups? Determine the current level of knowledge and background of the group. If the group has special problems such as physical or emotional handicaps, this will need to be considered when preparing the program.

Determine what kinds of materials, equipment and human resources are required to produce and present the program. At this point, a list of supplies, equipment and people necessary to produce and present the program must be compiled. A partial list might include projector, screen bulb, extension cord, camera, film, processing art work, titles and such. Determine what equipment is available and what will need to be secured. At this point, determine from what sources financial support can be obtained. The size of the budget for the program may affect its scope. The people who will help produce the program are important considerations. If a firefighter on the department has photography as a hobby and has a home darkroom, this may greatly reduce the costs. He could take all pictures and then develop the slides in his darkroom for much less cost than having them produced commercially.

The length and degree of sophistication of the program must be considered. How many minutes will it take to view the program? The length of the program may be dependent on the audience for which it is intended. Preschool children have shorter attention spans than older children and a program for them will necessarily have to be shorter. The finances of the program and talent available may also influence how elaborate and complex the program will be.

If the physical facilities in which the program will be shown and utilized are known, this may influence production. Once the above factors have been considered and decisions made, the actual production process can begin.

Perhaps the most effective way to begin planning an audio-visual program is through the process of "brainstorming." Collect all your ideas for the project by writing them on 3- × 5-inch cards. Use only one idea per card. Then consider each of the ideas separately, using the questions what, where, when, who, why and how. Write the responses to the questions on the card.

It is highly recommended that ideas and suggestions from others be enlisted in this process of brainstorming. Write all ideas on the cards — even those which do not seem feasible. Do not discard an idea merely because it did not generate from your own thinking session. Ideas from other sources may prove to be most pertinent in contributing to the success of the program.

Once all the ideas have been collected, begin arranging the cards in a logical order. At this point you may discover that some ideas may need to be added or deleted in order to make the concept flow smoothly. When the sequence of ideas follows in a satisfactory pattern, number the cards consecutively to retain the order.

At this point it is desirable to determine how each idea may be illustrated. Draw a rough sketch of each concept on the various cards, imagining how the camera might capture each idea. Sketch each scene as you wish it to progress; there may be as many as forty pictures to complete a total idea. This will make it possible for the person shooting the pictures to comprehend what the camera must convey. The sketches arranged in sequences are called a "storyboard."

A storyboard will aid the photographer in taking the pictures that are desired for the slide program.

Before attempting to photograph the project, certain concepts should be understood in relation to conveying ideas to an audience. That which is stimulating and interesting will capture the attention of the viewers more quickly than that which is lifeless and dull.

- *Motion* — People appreciate action. Slides should not stay on the screen more than 15 seconds.

- *Intensity* — Loud noises, brilliant lights and intense color may be manipulated to attract attention.

- *Size* — Large signs, big headlines and camera close-up views can be used to demonstrate a point.

- *Contrast* — Large beside small; light next to dark; fast contrasted with slow; loud, then soft — all contribute to creating contrasts.

- *Repetition* — Slogans and key phrases used repeatedly will instill the viewer with a recall of the purpose of the program.

In addition to the previously mentioned devices, there are three terms which should be defined which will have an infuence on how successfully the program will be assembled and presented for the audience.

> Attitudes — A way of looking at things. Characteristic way of acting or feeling toward a given situation, person or thing.
>
> Conditioning — A response that may be said to be instinctive (pulling one's hand away from a hot pan).
>
> Habit — A fixed way of responding to a given situation.

It may be necessary to change the public's attitudes and habits toward fire safety

Attitudes and habits can be changed, resulting in conditioned responses. Before we can expect people to change their attitudes concerning fire safety, we must demonstrate to them in what ways it will be to their advantage to do so. That is the primary purpose of educating the public in fire prevention.

Keeping these concepts in mind, refine the idea cards to give specific details regarding size, color, repetition, intensity and motion. The commentary on these cards may be only a word, phrase or single sentence. It may be part of the final script.

The planning cards will provide the basic information for creating the script. When working on the script, divide the paper into two separate columns. Use one column for the slide or visual description, the other is for the narration. Once the slides are compiled and the narrative written in final form, decide if audio tapes are feasible. If audio is to be used, have a person who has clear, concise speech record the narrative.

Now you are on the way to creating a program that will help your community to have a fire safe future. The many hours that will go into the preparation will be rewarded many times over by the satisfaction of serving a great human concern.

Before the actual production of the program, consider what legal responsibilities are valid. Determine who will own the program and if the program will be copyrighted and insured. Make certain that there will be no copyright violation while producing the program. If copyrighted articles or recorded music are to be used, gain permission first.

Make certain the program produced will not violate any copyrights

After the program is produced, pilot test it for critique and evaluation. Determine if the program has the desired effect on the audience and if the objectives are met. Determine if the audience will be concerned about a specific problem, can perform a task, demonstrate a skill or take proper action. If the program has weak parts, correct them at this point before it is released for general usage.

FIVE-SCREEN SLIDE PROGRAM

This type of program is becoming extremely popular because of its effectiveness in combining sound with multiple, variable visuals. A 20-minute program using five screens requires approximately 80 slides per tray. This means that over 400 slides are necessary to obtain optimum results. When utilizing this type program, allow adequate time to set up before the actual presentation begins. It will require between 30 and 45 minutes for two people to assemble the equipment and assure that all components are working properly.

Five projectors mounted on projector table. A tape recording can be used for the audio portion of the program. Music used in conjunction with narrative creates a good effect. It is the responsibility of the person controlling the projectors to coordinate the slides with the taped audio.

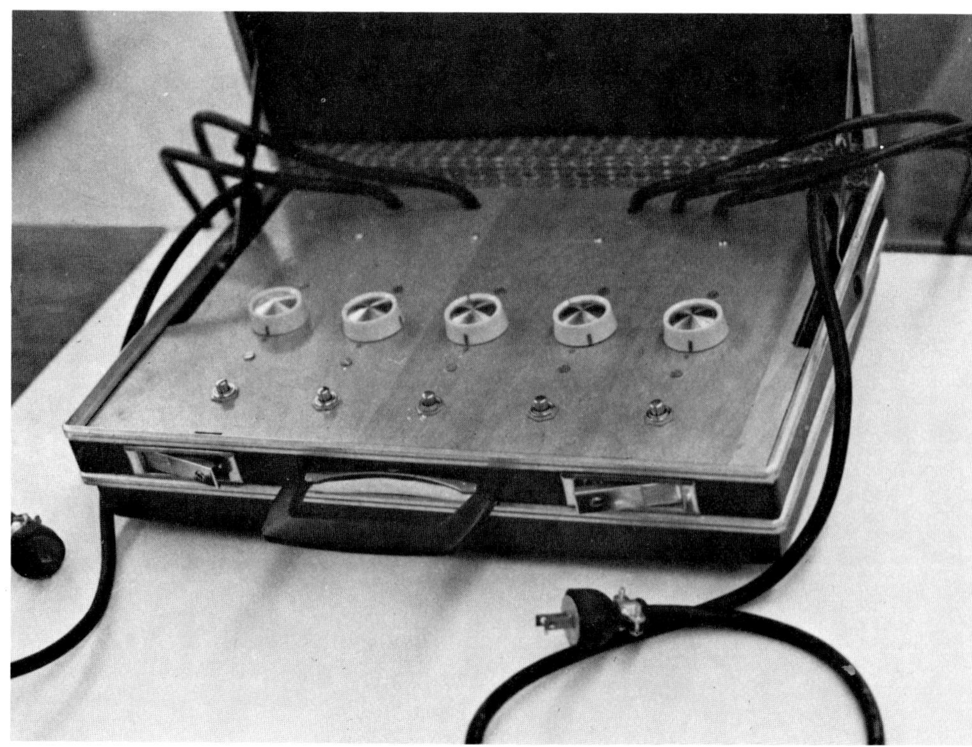

The controls for this assembly are made of five refrigerator door switches and five 110 volt dimmer switches. Notice that the entire control panel is enclosed within a briefcase for good protection and easy transportation.

These plugs come out of the control panel and connect with the five projectors. Color coding or numbering at each end will help coordinate proper hook-up.

Extensions for the legs of the projector table allow for controlling the height of the table. A screw in the table legs can be loosened for adjusting the height of the table. Tightening the screw puts tension against the inner tube and holds the position.

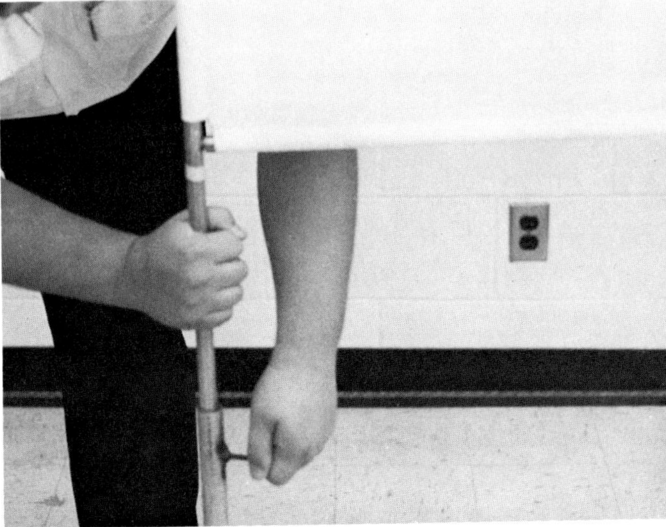

The economical home-made screens must be assembled to continue the preparations. An expansion fitting on the end of the ½-inch horizontal tube allows the screen to be stretched tight.

Other set screws in the screen base legs allow for adjusting the height of the screens in the same way as the table was adjusted. Being able to adjust either screen or table gives best flexibility for a variety of situations. The program can also be modified to use more or fewer projectors and screens.

TABLE TOP DEMONSTRATIONS

Fire has always fascinated man, and wise educators have harnessed this mystic to hold the public's attention during presentations. Several demonstrations throughout a program hold the audience's attention and aid in message retention. In order to be effective, the demonstration must move along and not dwell on one subject too long. Table top demonstrations are excellent visual aids for small groups.

Determine what is to be shown during the demonstration, what materials will be needed and to whom the demonstration will be given. Some basic rules to follow for presenting effective demonstrations include:

- The presentation must be interesting and polished. Demonstrations should not be used to wake people up during a slow or boring presentation.

- The demonstrations must support the material being presented. If flammable liquid hazards are being discussed, the demonstration must deal with flammable liquids.

- Demonstrations are intended to support the presentation and not to be a substitute for the presentation. All demonstration — with little or no presentation — resembles a magic show more than a public education presentation.

Demonstrations utilizing more complex equipment can be constructed by most fire departments.

Demonstrations can be very simple such as this one utilizing a skillet, a lid and rubbing alcohol. Regardless of the equipment and materials used, be sure the demonstration can be easily understood by the audience and properly complements the presentation.

- The demonstration must be on the same level as the audience. Different kits are needed for different age groups.

To assemble a kit, the first step is to develop a well-rounded, effective and smooth presentation. The presentation should be technically correct and interesting. To determine if the presentation is ready for added demonstrations ask: "Would the presentation be effective without demonstrations?" If not, it probably will not be effective with demonstrations.

When the presentation is complete, review the main topics and select the major points for highlighting. These major points are then developed into demonstrations. The criteria for a demonstration include:

- It must be safe. Although a demonstration should be dramatic and impressive, always consider the safety aspect first.

- It must adequately demonstrate the point. Gear the demonstration to the point being made.

- It must be visible. If people cannot view the demonstration, the effectiveness is lost.

- It must be simple and easily understood. Demonstrations that are too technical or complex may not keep the audience's attention.

Guidelines for effective demonstrations

When the demonstration has been selected, the components are gathered and the display assembled. The demonstration should be set up and practiced several times, while varying the conditions and amount of fuel. The educator must be familiar with the operation of the demonstration and well prepared for any reaction which might occur. Know the subject well enough to answer any questions that might be posed.

After all the parts have been gathered, a box can be designed to carry the kit. Generally it should be designed on paper prior to being built to avoid some problems that might be overlooked during the initial stages.

When the kit is complete the educator should practice setting up and taking down the demonstrations in the station. The more practice he has the easier it will be to set up, and the less likely he will be to leave parts of his demonstration behind.

Some common demonstrations found in fire service presentations are discussed. These demonstrations can be easily assembled at relatively little expense.

AEROSOL FLAMMABILITY

The hazard being demonstrated is the use of aerosol products around open flames or while smoking. Several different aerosol products, a butane lighter and a glove are required for this demonstration. A variety of aerosol products such as insect sprays, household cleaners and hair spray should be chosen. It is recommended that the aerosols be tested prior to the demonstration to assure that they are flammable and are still under adequate pressure for effective use in the demonstration.

To demonstrate the flammability, hold the butane lighter in the gloved hand well away from the body and light a tall flame. Hold the aerosol in the opposite hand and spray the contents above the flame. The bottom edge of the spray pattern should touch the tip of the flame and a large fireball should occur.

CANDLE AND JAR

The principle demonstrated in this presentation is the extinguishment of flame by oxygen depletion. This should be used to explain how fire can deplete the oxygen in an enclosed area and thus cause asphyxiation in humans. The objects needed for the exhibit are a small heating candle such as those used for fondue pots, a quart or liter jar and a butane lighter.

Light the candle and allow it to burn momentarily while the oxygen level of normal air is explained. Place the jar over the candle and the flame will soon dim and extinguish. This represents oxygen depletion to levels below that necessary for flame or life.

CANDLE BOX

The point being made in this demonstration is that air containing more oxygen can be found closer to the floor in a fire situation. This is the reason that fire service personnel emphasize crawling when in a smoky or fire atmosphere. A candle box and seven heating candles are required for this demonstration. To construct a candle box, the materials needed include: one-inch by one-inch lumber, two glass panes (one for the front and one for the back), one-inch by four-inch lumber, suitable nails and paint. The lengths and sizes of the lumber will depend on the size of box desired as will the size of the glass panes.

To assemble the box, build a three-sided stand with a removable top out of the one- by four-inch lumber. The inside facing of the stand is channeled approximately one-half inch from the outside edge to allow the glass to be slid in and out easily. The one- by one-inch lumber is then cut into pieces to form a stairstep for the candles to set on. The box is then assembled and painted. If a person with artistic ability is available, a man crawling and a man standing can be painted on the front of the glass between the candle locations.

For the demonstration, remove the back glass and light the candles. Replace the back glass and top and the candles will extinguish from the top down. The candles at the level of the man standing will go out long before those at the level of the man crawling. This demonstrates visually the importance of staying low during a fire.

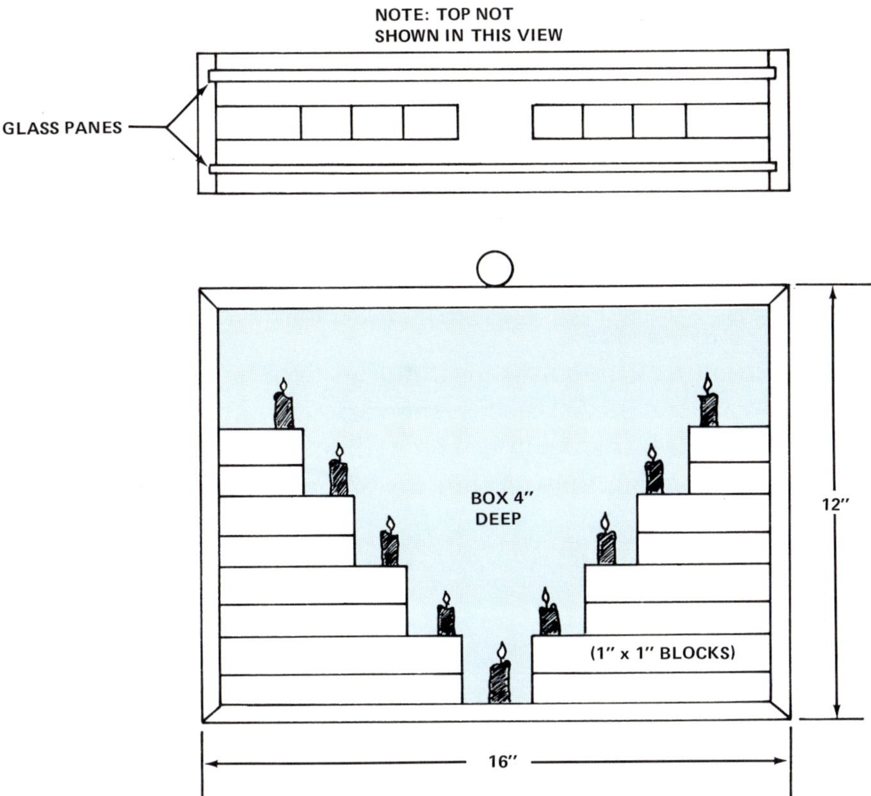

A candle box can be used to visually demonstrate oxygen depletion. It can also be used to show the extinguishing effect of an inert gas such as carbon dioxide.

CANDLE BOX CO_2

The purpose of this demonstration is to illustrate the smothering and extinguishing effect of carbon dioxide (CO_2). This demonstration requires the same candle box as previously described and a container of CO_2 gas. A gallon or four-liter glass jar filled from a CO_2 extinguisher and sealed works well. Label the jar with the designation CO_2 in blue letters. The blue lettering indicates an inert gas.

Light the candles but do not replace the cover. Allow the candles to burn briefly and then pour the CO_2 onto one end of the stairs. The candles that the CO_2 passes over on its way down the stairs will go out. Shortly after the first candle goes out, stop pouring and let the CO_2 settle in the bottom of the box. The bottom candle will then go out. After indicating the approximate level of CO_2 in the box, pour a little more CO_2 in the box. The next higher candle should go out. Finally, fill the box with CO_2 and the remaining candles will extinguish.

ILLUMINATED FIRE TRIANGLE

The intent of this display is to emphasize that all three sides of the fire triangle are necessary for fire to exist. To create this display requires: three double-terminaled, two-poled switches; four light receptacles; four 15-watt light bulbs — one each of red, blue, yellow and clear; an adequate amount of wire; a plug; a plywood triangle with a minimum of two-foot sides; an opaque plexiglass triangle equal in size to the plywood triangle; and two-inch by one-half-inch stripping.

Assemble the triangle according to the diagram and place the three switches in the back of the plywood. Wire two circuits to each switch with one circuit from each switch going directly to one of the lights on the outside of the triangle. The second circuit is wired in series across all three switches to the light in the middle. Letter the plexiglass with the words FUEL, OXYGEN and HEAT along the outside edges. Letter the word FIRE in the center.

Turn on all three switches and all four lights will come on indicating a fire. As you discuss the various components of the fire triangle, turn off that switch. The light on that side and the FIRE light will go out simulating that the fire is out.

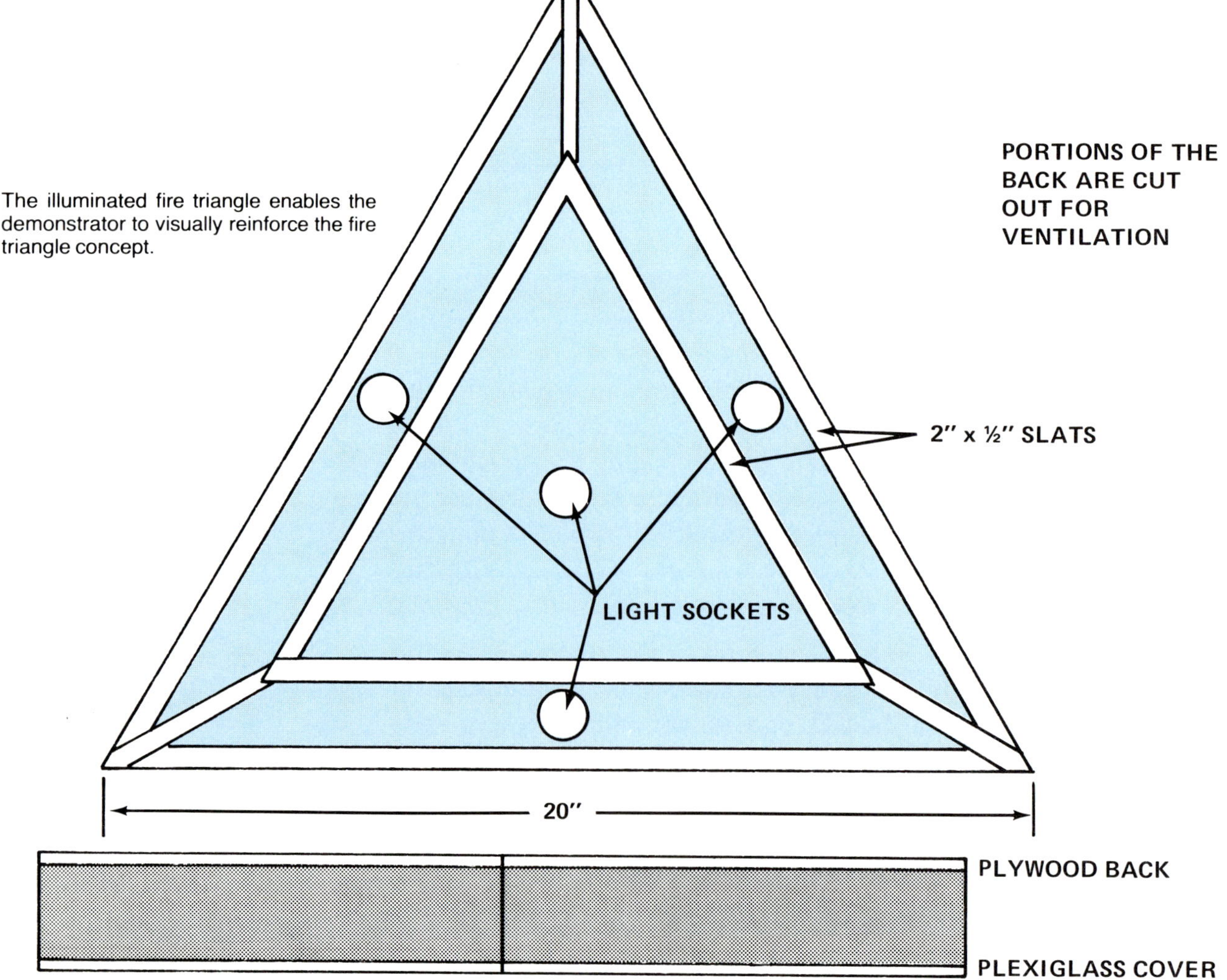

The illuminated fire triangle enables the demonstrator to visually reinforce the fire triangle concept.

FLAMEPROOFING

The objective of this demonstration is to show that flameproofing material provides a definite life safety advantage. Equipment and material needed for the demonstration are: two similar Bunsen burners; adequate gas hose with a tee; propane bottle; striker; a ring stand with a double cross arm; two clamps; and one-inch by six-inch strips of both flameproof and non-flameproof material.

To begin the demonstration, position the cross arm on the ring stand approximately seven inches above the height of the Bunsen burners. A sample of both materials is attached on opposite sides of the cross arm. The ring stand is moved away from the burners, the burners are lit and adjusted to equal flame size. Move the samples over the flames and note the difference in burn characteristics. Impress on parents the need for purchasing flameproof sleepwear and Halloween costumes for their children.

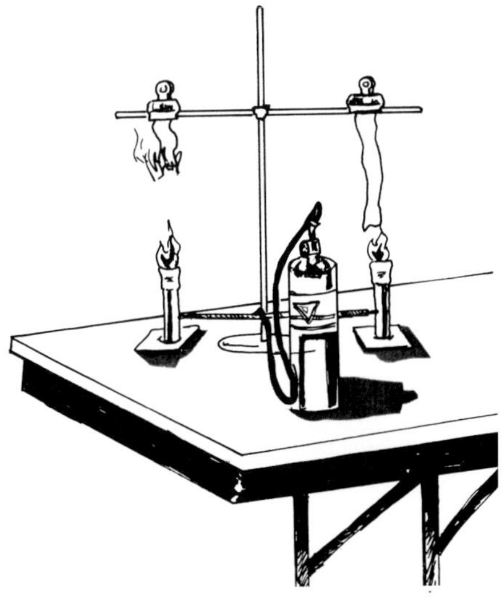

The merits of flameproof material are adequately shown in this demonstration.

EXTINGUISHMENT WITH LID

The concept being demonstrated is that most kitchen grease fires in frying pans can be be extinguished by simply covering with a lid. Stress the importance of always having a lid that fits the pan readily available. The materials required for this presentation include a small frying pan with a suitable lid, rubbing alcohol and a butane lighter.

Fill the bottom of the frying pan with alcohol, light it and allow the alcohol to burn briefly. Slide the lid over the pan from the side and bring it to rest on the pan. After a short period, remove the lid to reveal that the fire is out.

BAKING SODA EXTINGUISHMENT

This demonstration simulates how household baking soda can be used to extinguish grease fires. This demonstration is especially applicable when talking to women's groups. Rubbing alcohol, baking soda, a flour sifter, a small pan and a butane lighter are required to demonstrate this extinguishing method.

Put a small amount of alcohol in the pan and the baking soda in the sifter. Light the alcohol and allow it to preburn for a brief period. Dispense the baking soda onto the alcohol using the sifter and the fire will be extinguished.

SMOKEHOUSE

This demonstration shows that smoke first travels upwards in a house and then is pushed downward as the upper area becomes saturated. This demonstration again illustrates why it is necessary to crawl during fires. A two-story doll house with the windows sealed can be utilized. Cover the front with airtight plexiglass. Openings at the top and bottom of the house are necessary.

Place a smoke candle in an appropriate container, light it and set it in the house. The smoke can be observed as it first fills the top of the house and then continues to spread downward until the bottom is filled.

DUST EXPLOSIONS

This demonstration illustrates the high propagation characteristics of finely divided combustible dusts. The demonstration requires the construction of a dust explosion box as illustrated below. Construct the box with dimensions of 22 inches in length, 10 inches high and 8 inches deep. Cover one end with a light metal door hinged at the top. A glass front is put on the front so that the demonstration can be easily seen. A dust tray with an air line is built on the top center of the end opposite the door.

Put approximately one tablespoon of combustible dust such as powdered sugar or cornstarch on the dust tray. A source of ignition such as a candle or a set of arcing electrodes is placed inside the box. The dust is dissipated into the air by blowing through the hose. When blowing, one quick forceful breath should be given and then the hose removed from the mouth. Shortly, the dust will flash, blowing the door open.

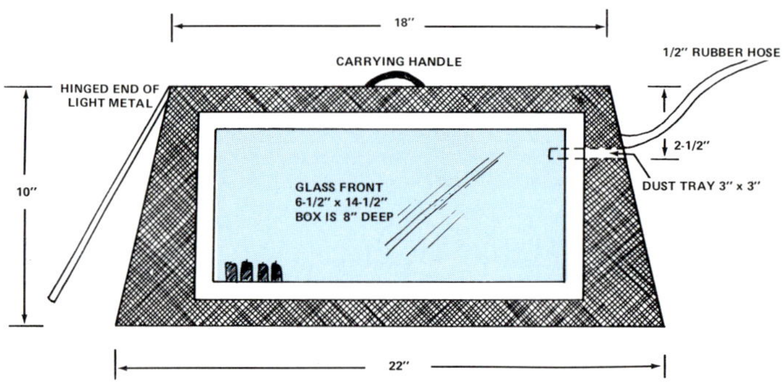

CAUTION: DO NOT USE ORDINARY GLASS.

The flammability of finely divided dust is shown by utilizing a dust explosion box.

VAPOR TRAIL TUBE

The concept demonstrated is the movement of flammable vapors downhill. The material required includes: a four- to six-foot, one-inch diameter plexiglass tube; two mounting stands; a six-inch length of one-inch diameter metal tube; a one-half inch tea ball filled with cotton; an eyedropper of gasoline; a candle and a butane lighter. Cut the metal tube lengthwise down the center for four inches. Cut one side perpendicularly to the length to remove one-half of the tube. Insert the uncut end of the tube into the plexiglass tube and secure by drilling through the tubes and bolting. Attach the tea ball to the open end with a screw. Attach the mounting stands to slant the tube at an approximate 30° angle with the metal at the high end. The open side of the tube should be up and the low end should be high enough for the candle to sit underneath.

Light the candle and put a dropper full of gasoline on the tea ball. The vapors will travel down the tube to the candle. The flash up the tube to the tea ball will be visible and the tea ball will burn until extinguished.

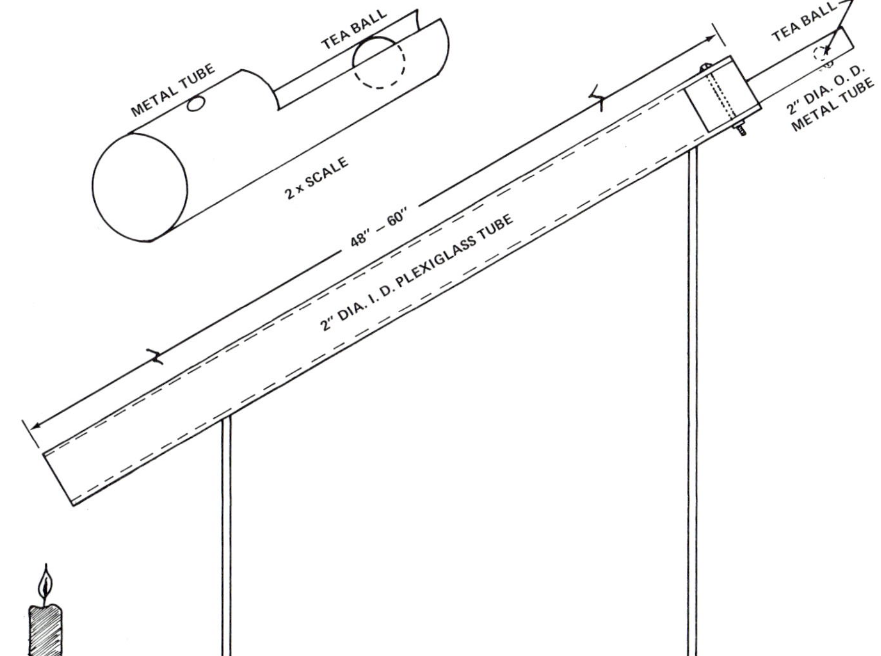

The vapor tube demonstrates to the audience the manner in which a flammable vapor can travel invisibly to an ignition source.

VAPOR EXPLOSION TUBE

This demonstration illustrates the tremendous energy stored in gasoline and the potential damage it can do. This demonstration requires: an eight-inch plexiglass tube two inches in diameter; two 1/4-inch plexiglass base plates, six inches by six inches; two 3/16-inch bolts, 1½ inches long; four nuts; a small amount of gasoline and an eyedropper; a twelve-volt car battery; a twelve-volt coil; two alligator clips; one switch; and adequate wire.

Drill a 3/16-inch hole in the tube approximately two inches from the bottom. Place one nut on each bolt next to the head. Thread the bolts through the plexiglass tube with the heads inside and adjust to provide less than a 1/16-inch gap between the bolt heads. Put a nut on the outside of the bolt, and tighten the two nuts against the plexiglass to secure the bolts in position. Cut out one base plate to allow the plexiglass tube to fit snuggly through. Glue the two base plates together and glue the tube into the cut out hole. The primary circuit of the coil is wired to the battery through the switch. Connect the secondary coil leads to the two wires that are attached to the alligator clips.

Place two or three drops of gasoline in the inside of the tube and allow it to vaporize. Attach the alligator clips to the ends of the bolts and turn the switch on. A loud bang will ensue.

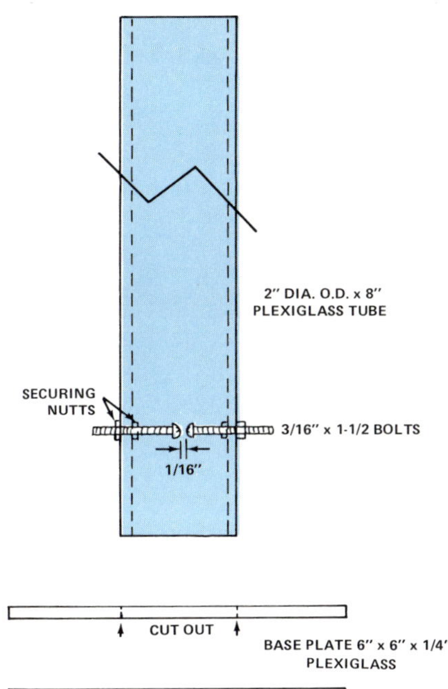

The gas explosion tube will quickly gain the audience's attention as it demonstrates the power of a few drops of flammable liquid.

DEMONSTRATION KITS

A demonstration kit to illustrate several points during a fire prevention program can be assembled at relatively little expense. A kit such as this one assembled by Joe Day of Los Angeles County provides an orderly and compact carrying case for all needed components for a particular presentation. The advantages of a kit are that the orderliness of the equipment lessens set-up time and assures that all materials required are available.

A wooden carrying case with handle allows for easy transportation. A case such as the one pictured here can be built at the fire station.

This particular kit includes items for use in several demonstrations. The tube in the top of the kit is a vapor travel tube used to illustrate how vapors can travel invisibly and then flash back from a distant ignition source. Behind the tube is a flameproof drop cloth which protects the presentation table top from hot objects. In the upper right corner aerosols are stored to demonstrate the surprising flammability of some common items.

The middle left compartment houses a plexiglass dust explosion chamber. Powder is placed in the funnel, air is then injected into the bottom of the funnel and the powder is ignited by an arc across the electrodes. The leads for the electrodes are stored to the left of the dust chamber. The center middle compartment is used to store demonstration supplies such as clean-up equipment, powders for the dust chamber and flammable liquids. Note the use of the UL-listed safety can for flammable liquid storage.

Shoe polish, (a flammable solid), is stored in the lower left compartment and is used to demonstrate the flammability of common household substances. To the right of the shoe polish is a ring stand used to elevate certain demonstrations for easier viewing and added emphasis. In the middle bottom compartment a small frying pan with a lid is provided to show one good method of putting out kitchen pan fires. Next to the pan is a commercial electrical arc generator used to produce ignition arcs for the dust chamber and vapor explosion tube to the right. The vapor explosion tube is made of plexiglass and used to demonstrate the energy of a small amount of flammable liquid. The extension cord is utilized to get power to the arc generator.

Some of the items used in this demonstration kit such as the electrical arc generator have been commercially purchased. It is possible, however, to design and build these type items if finances are limited.

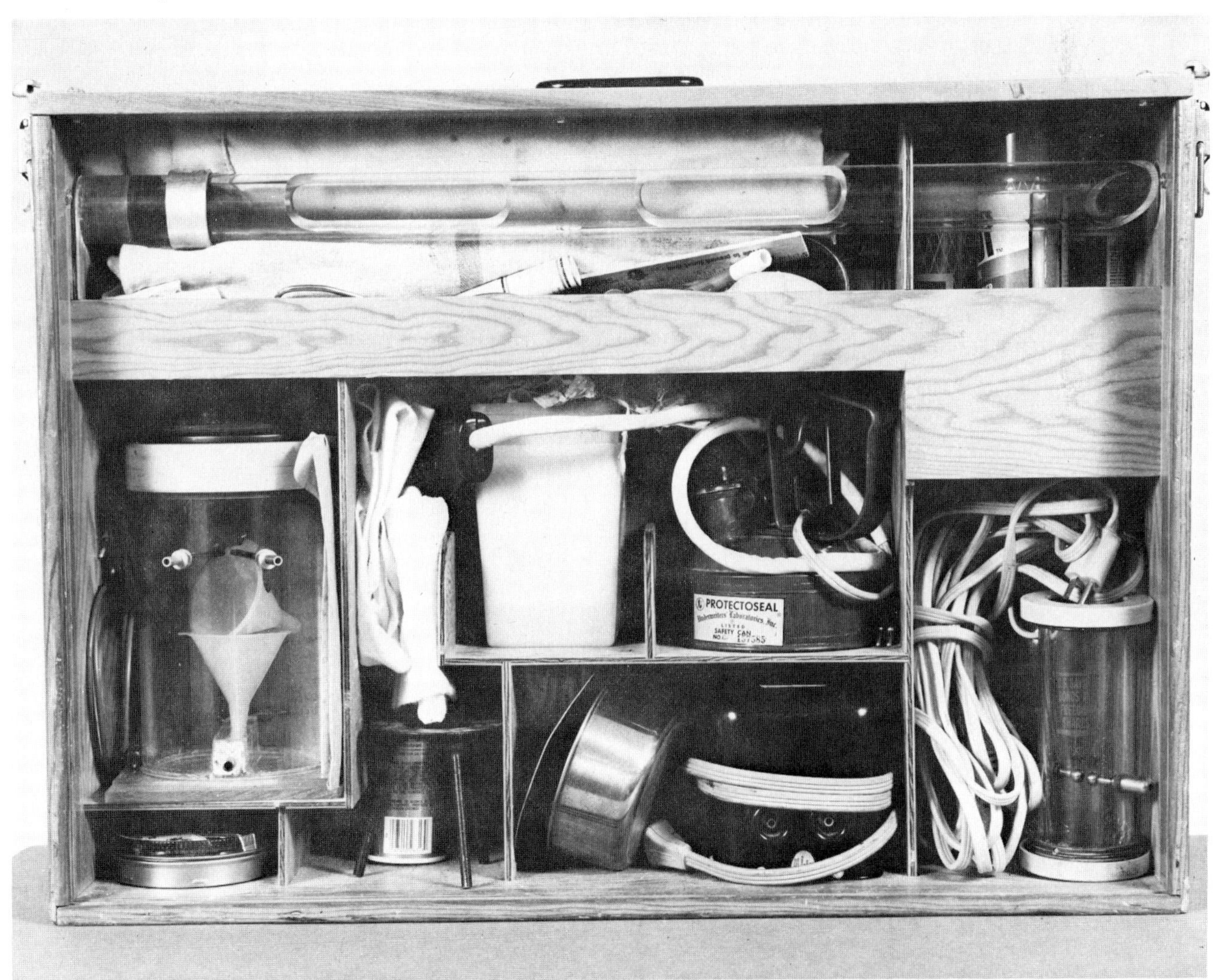

A DEMONSTRATION EXAMPLE

Topic: Hazardous flammables common to the home.
List of materials: paint spray cans, furniture polish, hair spray, drain cleaners, bug spray, fire extinguisher and metal can with lid.

In your home there are a host of flammable materials that are common to almost every home in America. (Hold up some of the products.) These products can be used safely, but extreme care must be exercised. Be sure to read the label, especially the caution portion of the label.	*Hold up a can and point to the caution label.*
Caution: Keep out of reach of children, contents under pressure. See back panel for additional cautions.	*Read the label of a can or two.*
(From back panel) Caution: harmful if swallowed. Do not puncture or incinerate. Do not spray into open flame or near fire. Exposure to heat may cause bursting. Avoid prolonged exposure to direct sunlight. In case of skin contact wash thoroughly with soap and water. Extreme care should be taken with the above product. The label indicates that the product may have something in it that might cause damage to the skin or be harmful if swallowed or inhaled. It also might burn if sprayed into flame or near fire. Also, prolonged exposure to the sun or heat may cause the can to burst.	*Hold up a dust control product.*
Let's just see if it will burn (blow torch effect). How about that?	*Spray contents of can into flame from a match.*
Now let's take a small dust rag without this product sprayed onto it and see how easy it catches fire.	*Take a match to an untreated rag.*

Now treat the same rag with the dust control product. There is no label on the rag to describe its danger or proper disposal procedures. There is nothing to say that if the baby gets hold of the rag and puts it into its mouth, it could get very sick. The caution label says to keep out of the reach of children, but it *assumes* that the user will take the same precautions with rags and mops. What if a baby crawled into a closet with a treated mop and put it into his mouth?

Take a match to the treated product. (Note the difference.)

Nothing on this label tells what is inside. There is no way for the consumer to know what the product contains or what the propellent is. Disregarding the product, let's look at the propellant. Freon, according to recent findings, destroys the ozone layer protecting the earth's atmosphere, so a big push by environmentalists has managed to get it reduced as a propellant for aerosols. What is considered for a replacement? LP gas without an odorant added.

Show the label again.

Let's look at some other products in the home, such as this bug killer. Look at the label. "DO NOT SPRAY NEAR OPEN FLAME. DO NOT SMOKE WHILE USING. CAUTION: COMBUSTIBLE MIXTURE. Active ingredients: almost 90% petroleum distillate. 2-(1-methyletloxy) pheno methylcarbamate 0.665%; -2,2 dichlorovinyl dimethyl prosphate (DDVP) 0.186% and 0.014% related compounds; petroleum distillate 89.1%. Harmful if swallowed, inhaled or absorbed through the skin. Provide adequate ventilation in average room."

Hold up a can of liquid ant and roach killer.

When the label says it is a combustible mixture, extreme care should be maintained at all times.

Spray this liquid through a flame.

When sprayed on an object, it makes the object much more flammable. If the object is porous and absorbs the petroleum product, it could remain flammable for some time after it appears dry.

Spray onto a piece of cardboard.

ELECTRICAL CORDS - FLANNEL BOARD STORY

A flannel board story such as this one can be relatively simple and inexpensive to create. Yet, it is very effective when working with small children. Remember that children's attention span is relatively short so create the program accordingly. Visuals combined with participation are a good way to teach children about fire safety. This program, created by Oklahoma Fire Service Training, combines both visuals and participation to instruct children on the potential hazards of electrical cords.

OBJECTIVE

The objective of this program is to enable small children to recognize electrical cords in association with electrical appliances found in their environment. It is also to teach the child to recognize the danger associated with electrical cords.

MATERIALS NEEDED

- Flannel board
- Graphic pictures of electrical objects
    T.V.
    Radio
    Clock
    Mixer
    Coffee pot
    Stereo
    Blender
    Electric blanket

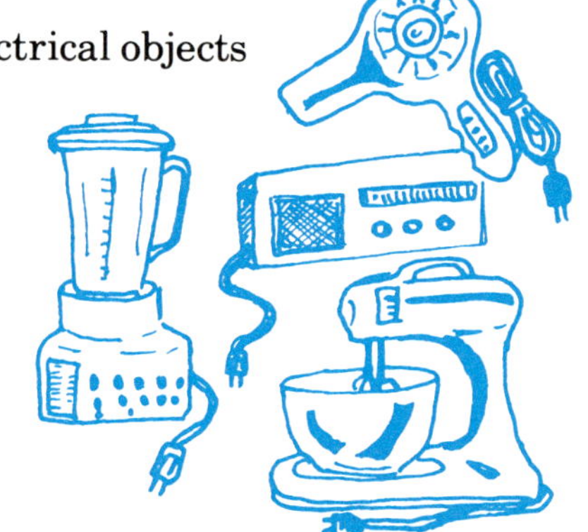

Curling iron
Table lamp
Electric skillet
Refrigerator
Blow hair dryer
Electric rollers

- Yarn
- Small lamp with low intensity bulb (non-frosted)
- Capital E flashcard

PROCEDURE

Introduction

General greeting

Enlist help of children

Give each child an object.

What picture do you have?

Children can place objects on flannel board one by one. As each one places object on board ask these questions.

Do you have a . . . . at your house?
Where is the . . . . in your house?
    (location of object)

Place a length of yarn from the object

Does it have an electrical cord?

When all of the objects are on the board ask these questions.

What makes each of these work?

Do they have an electrical cord?

What turns these things on?

Hold up the big E flashcard

Electricity is a good E word.

Electricity makes all of these things (name several objects) *work.*

Let's all say electricity together. It's a fun word to say.

After a child has gained a basic understanding of the electrical cord, the educator should then talk about electricity while using the lamp as a visual aid.

Let's look at this lamp. Does it have a cord? Does electricity make it work? How can we make it work? (plug in cord, turn on light) *We can see the electricity in this light bulb. There is electricity in this cord too. Can we see the electricity in the cord? No, we cannot see the electricity in the cord, but it is still there. Electricity travels from the outlet to the bulb in this cord. We can touch the electrical cord and the electricity will not hurt us. It is inside (in the middle of) the cord. The electricity can hurt us if we let it out. How can the electricity get out and hurt us? The rule about electrical cords is — children don't play with cords. Do you have a baby brother or sister at your house? An electrical cord would hurt them too. Babies always put things in their mouths — this is when an electrical cord would really hurt. Help keep cords away from babies.*

CONCLUSION

We have talked about electricity today. Let's say it together one more time. Remember there is electricity inside cords and if it gets out it will hurt us. Electrical cords aren't to play with.

THE STORY OF THE LITTLE RED FIRE HAT

A Fire Prevention Flannel Board Story for Young Children

This story was created by Nancy Dennis Trench, Fire Service Training, Oklahoma State University. This is an example of using visuals to make a presentation more interesting and effective. The visuals used can be produced at minimal expense and take relatively little time to make. Examine existing programs that do not incorporate visuals and determine if, by adding simple visuals, the program could be more effective.

The Story of the Little Red Fire Hat is designed to be presented to young children three to six years old. The flannel board works well on the floor with the children seated in front (remember a small group, maximum of 15 works best). For best results, tell the story in your own words with lots of expression. Place each hat on the board in sequence as it is mentioned in the story (see illustration). When the story is finished all hats will be on the board and the Little Red Fire Hat will be smiling.

Instructions for Making Felt Hats

Materials: Instructo felt - glue - scissors - tracing paper

Draw the hat shapes and use them for patterns to cut felt shapes. You will want to draw eyes, mouths, hat bands and accent shadows to use for patterns. Glue the pieces together to make the hat. Further instructions and colors are listed with each hat original. The felt hats "stick" to the flannel board without any adhesive. Easy to use!

Flannel board and instructo felt should be available from the school supply store in your area. If not, the material is available from:

Thompson Book & Supply Co.
101 N. University Drive
Edmond, Oklahoma 73034
(405) 341-0201

Instructo Flannel Board #7 — $14.95 (Folding 24″ × 36″)
Instructor Felt #64 — $5.50 (9″ × 12″ Sheets)

(Prices subject to change)

THE STORY OF THE LITTLE RED FIRE HAT

Once upon a time there was a LITTLE RED FIRE HAT *(Place fire hat on board)* who just loved being a fire hat! *(Put happy mouth on hat)* He was so red and shiny. But, *(change to sad mouth)* when the alarm sounded in the fire station and the little red fire hat got on the fire engine the siren was so loud it hurt his ears and he wished he were a COWBOY HAT. *(Place cowboy hat on board)* When they got to the fire there was always lots of heat and smoke. It hurt the little red fire hat's nose and it burned his eyes. When this happened he wished he were a STOCKING CAP. *(Place stocking cap on board)* The firefighters always used water to spray out the fire and it made the little hat wet and in the winter time it made ice on his head. The little fire hat wished he were a STRAW HAT *(Place straw hat on board)* out sunning himself on the beach. The worst thing of all was when someone would loose their home or their car or the place where they worked because of fire. That made the little red fire hat especially sad and he wished he were a BASEBALL CAP *(Place baseball cap on board)* out playing third base and having a good time. Poor little red fire hat. But, what did make the little red fire hat happy *(Change to happy mouth)* was when boys and girls would visit the fire station and he could tell them ways to be safe with fire at their home. What do you think the little red fire hat told the children?

This gives the instructor a perfect introduction for discussion concerning any fire prevention topic. Stop, drop and roll, matches aren't for children, crawl to exit in smoke, reporting a fire, and hazard elimination are just a few which might be approached. Consider the age of the audience in determining the topics and length of discussion.

Photograph courtesy of Oklahoma State University *Outreach* Magazine.

Using the visuals and the story of The Little Red Fire Hat will gain children's attention. After the story is completed, talk to the children about some aspect of fire safety.

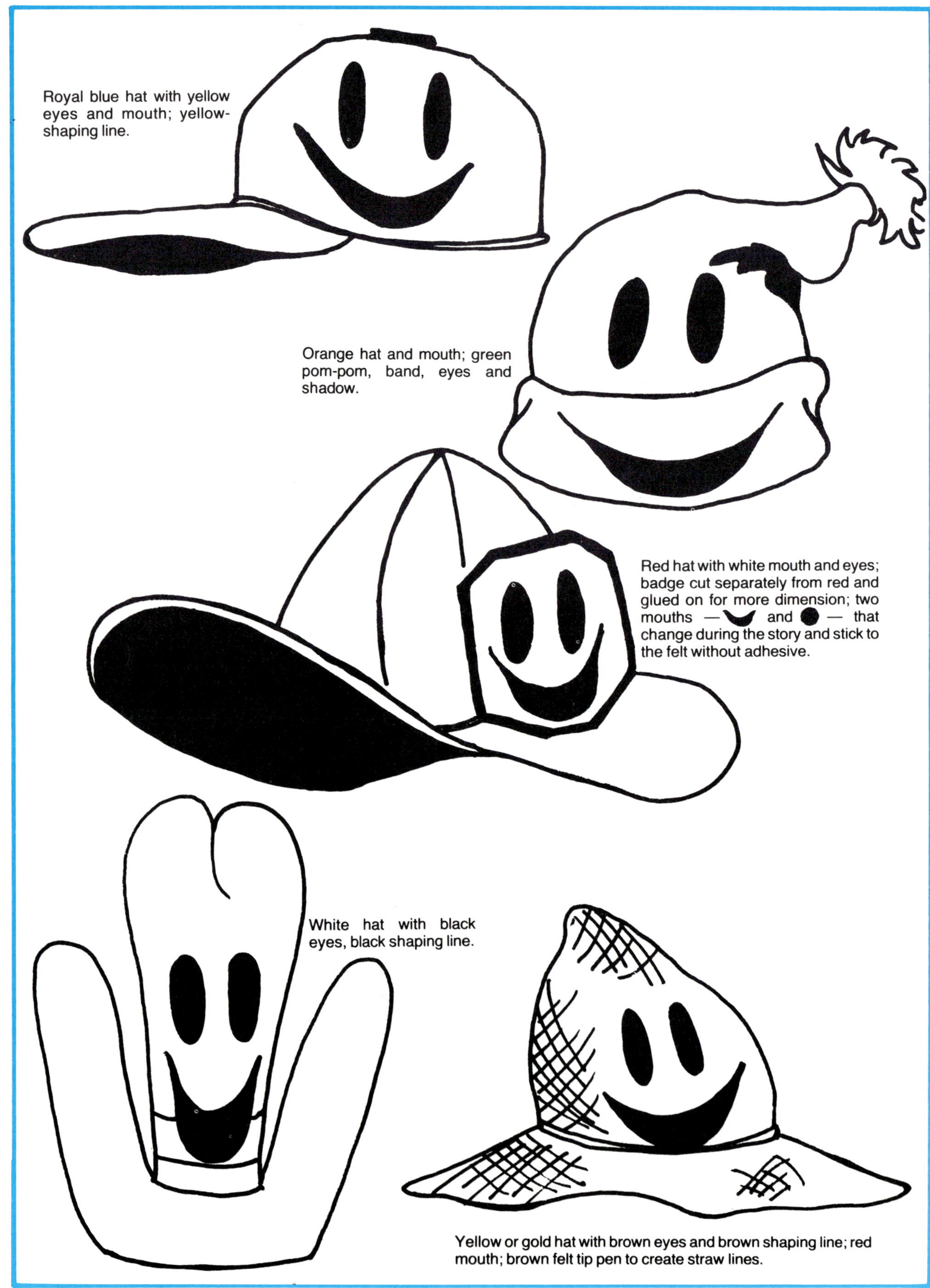
Royal blue hat with yellow eyes and mouth; yellow-shaping line.

Orange hat and mouth; green pom-pom, band, eyes and shadow.

Red hat with white mouth and eyes; badge cut separately from red and glued on for more dimension; two mouths — and — that change during the story and stick to the felt without adhesive.

White hat with black eyes, black shaping line.

Yellow or gold hat with brown eyes and brown shaping line; red mouth; brown felt tip pen to create straw lines.

Allowing the children to make their own fire hat to take home is fun and more strongly impresses them with the fire safety message that was discussed.

Photograph courtesy of Oklahoma State University *Outreach* Magazine.

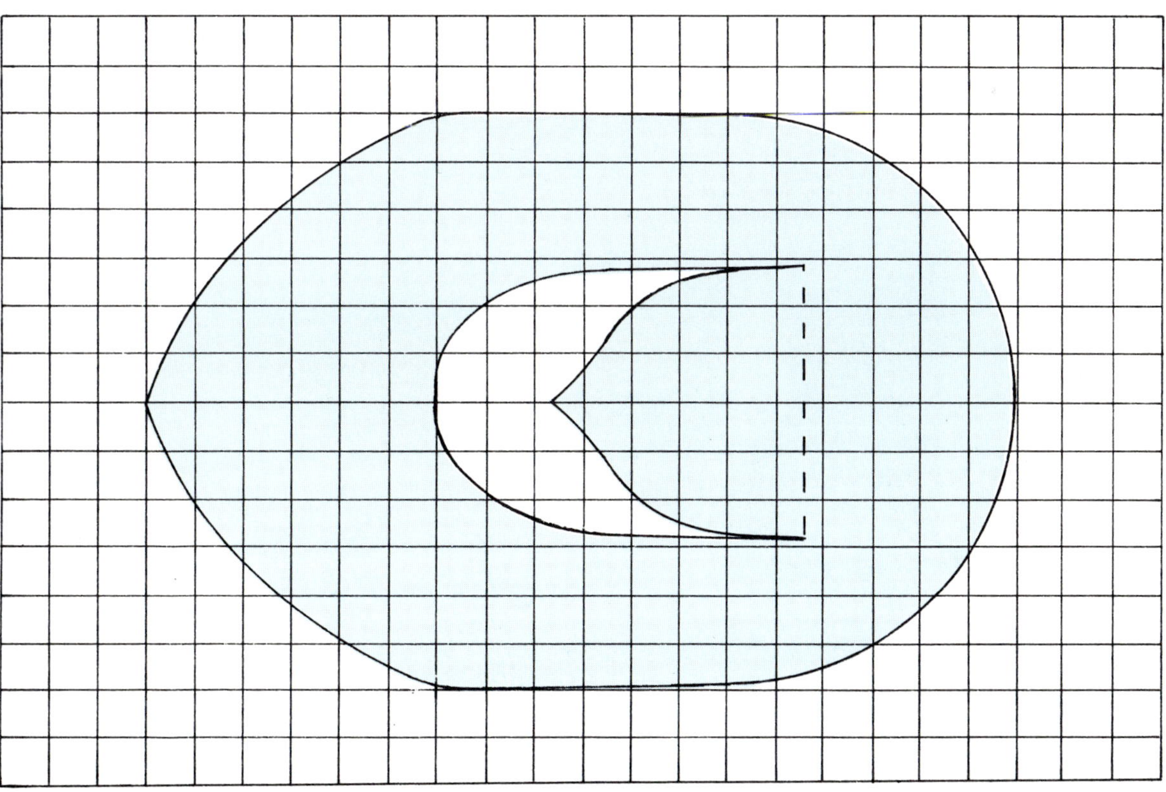

To make a red fire hat pattern, use this drawing which is scaled to where one block equals one inch. Use the cardboard pattern to trace hats for the children out of 12" × 18" construction paper. Fold along dotted line and place badge or gold star on front piece.

7
Resource Exchange

In public education, the words *convincing*, *encouraging*, *stimulating*, *involving*, *motivating*, *fostering* and *measuring* are common terminology among specialists in the field. We talk about *convincing* citizens there is a fire problem, *encouraging* an open and receptive attitude among skeptics, *stimulating* interests in prevention, *involving* the community, *motivating* participants, *fostering* joint efforts and *measuring* success. We're concerned with disinterest, ignorance and poor habits. We're dealing with behavioral change, commitment and responsibility; and since we're dealing with people, it takes **creativity**.

Creativity is limited only by the people involved. One person, no matter how brilliant, how innovative and how energetic, can accomplish only so much. If you multiply a single effort thousands of times with the creativity of others however, the potential is limitless. In other words, "two heads are better than one." Resource exchange is precisely that — building upon the strengths of the known and proven, drawing upon the expertise of others, sharing what is scarce and supporting one another. Each person involved in public education, whether veteran or novice, need not feel he must start from ground zero to accomplish anything.

A new trend is sweeping the country. Prevention is the key! Before, not after! Learn not to burn! People are doing wonderful things with public education. They have experienced failures as well as successes. Partial and complete programs have been developed and implemented. Numerous aids are available in the form of films, slide/tape/cassette packets, lesson plans, kits, printed pamphlets and brochures, recordings — the list is limited only by our creativity. The problem is how do we know what's out there, where do we get it, whom do we contact, what does it cost and what has been its success? Through resource exchange, answers to many of these questions are available for anyone interested in public education. The Public Education Office of the National Fire Administration has developed a multi-level resource exchange network to specifically deal with problems of this nature. Through this mechanism, public educators can gain access to information within a given community, statewide and nationally.

WHAT IS A RESOURCE?

A resource is something which can be shared. It may be intangible such as a theory on how to enlist the involvement of civic groups in a smoke detector campaign. It may be a film regarding children and matches, targeted for parents, which is loanable upon request. It may be an education specialist who has successfully designed and implemented a burn prevention program. These people are termed "resource persons" who can lend assistance by sharing a particular skill or know-how. A resource can be in printed form such as an elementary school safety curriculum, a pamphlet bearing a slogan or message, or a planning document. Status reports concerning research studies may be helpful and considered a resource. A listing of available grants or contracts is valuable and may be a resource for departments seeking funds to implement a project. In summary, a resource is anything or anyone which can impact on, contribute to or assist with the topic at hand.

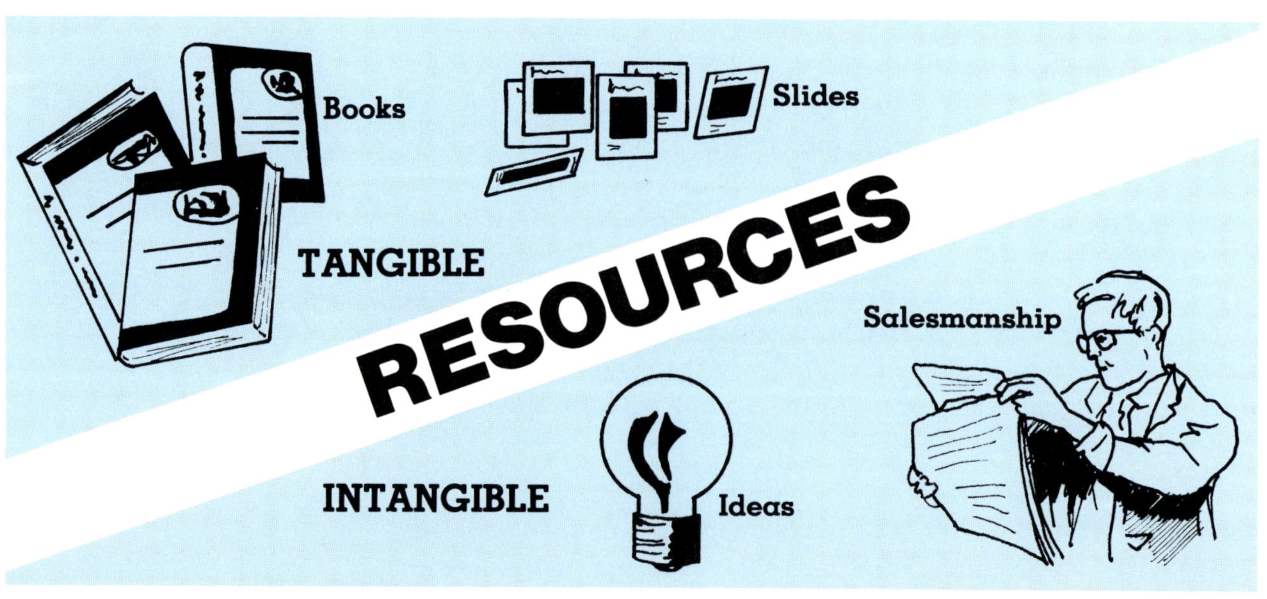

WHAT IS EXCHANGE?

Exchange is the actual sharing process. It is the transfer of information from one person to another. It may also be the movement of resource persons or materials from one place to another.

Exchange involves at least two parties — one to contribute and one to receive (user of the resource). Generally, it creates a growing effect. Once the initial transfer of information takes place, the receiver can assume a dual role as contributor and receiver, in turn sharing with another receiver. With each development in public education, exchange encourages sharing of accomplishments to spur others on as well as help make the job easier.

RESOURCE EXCHANGE SYSTEMS

So here are all these wonderful and perhaps not-so-wonderful ideas, programs, aids and special talents of people floating around in fire stations, schools, community businesses, homes, public assemblies, governmental offices and legislative chambers. The list of contributors and receivers is endless. We know what resources are and what exchange means, but how do we go about exchanging resources? Some mechanism to facilitate the sharing process seems desirable at this point. As the number of resources expands from a small diameter within a department to an enlarged circumference state- and nationwide, the need for a vehicle to promote information exchange becomes more evident. A Resource Exchange System is a mechanism structured to encourage the sharing of resources.

A Resource Exchange System enables individuals to share information and gain from each other's knowledge and skills. A person who is creating public education programs for the first time will find resource exchange invaluable.

There are people who have equipment, information or skills to offer and there are people who seek to use those things. The Resource Exchange System brings the two together to reap the benefits of the knowledge and accomplishments at hand. It exposes participants to successful activities, giving them a growing confidence that assistance is available, and creates a fraternal spirit in a mutual goal. It strives to eliminate duplication, "reinvention of the wheel," and other wasted efforts. The Resource Exchange System employs features which keep public education specialists informed, active and stimulated.

COMPONENTS OF THE SYSTEM

Resource Exchange Systems generally are comprised of four major components:

1. Information collection

2. Cataloguing and storage

3. Retrieval and dissemination

4. Feedback and maintenance

Obviously, some intake point must exist so that the resource can be centrally brought together. This is the information collection or gathering feature (Component 1), which permits contributors to feed information into the system. Contributions may be voluntary or solicited.

At the core of the system is an accumulation point where information is catalogued and stored (Component 2). Here review, evaluation, sorting by topic, indexing and storage of resources for future reference take place.

Retrieval and dissemination (Component 3) occur at the output point. The retrieval feature of the system deserves special attention during the design stage. Unless the data can be readily pulled out of storage and made accessible to the user, the system is seriously handicapped. Distribution of the information may be unsolicited or based upon specific request.

Lastly, a maintenance feature to continuously upgrade the quality of information provided by the system and to obtain feedback from participants is needed (Component 4).

To complete the circular pattern, keep in mind the one-to-one direct communications which are occurring constantly between fire educators.

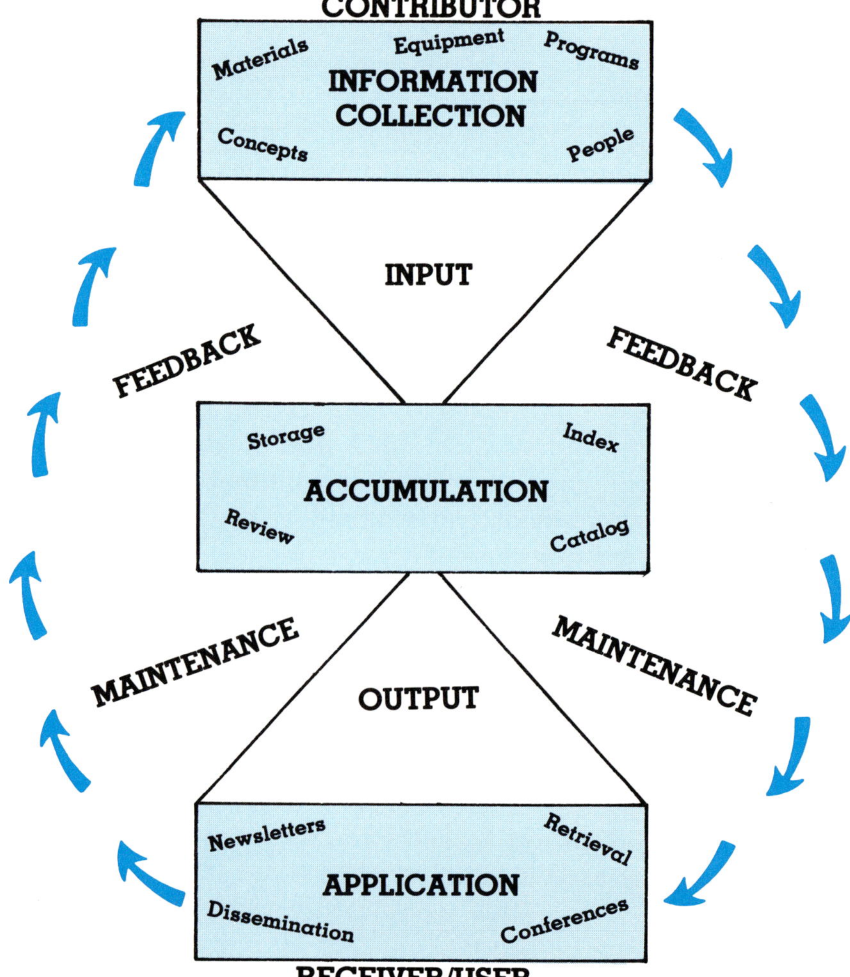

After information is collected, it is stored, reviewed and indexed at the accumulation point. The information desired by the user can then be retrieved and applied. Maintenance and feedback are continuous processes used to improve the quality of the information inputted into the system.

SYSTEM OPERATIONS

How are each of the four major components administered? What procedures are used to facilitate the sharing process? The selection and design of system features to accomplish each task will vary dependent upon the goals established, number of personnel involved and funding. Following are some of the more common methods used. Any one or a combination of several may be successfully employed in the system.

Information Collection

This process must be designed to permit easy input into the system. Usually a central contact is established and publicized. A question to be addressed before selecting collection features is: "Will information be actively sought for inclusion into the system or will input flow voluntarily from contributors?" It may be a combination of both, but the decision will have bearing on the processes chosen. Some common ways to gather resource information are:

- One-to-one verbal contact (Designate a contact person)

- Written communications (Designate an address)

- Staff meetings (Minutes often refer to activities which can be followed up for additional details)

- Newsletters and bulletins (Articles, announcements, and tidbits are extremely valuable)

- Manufacturer catalogues (A ready reference of what is available at a cost)

- Media, resource and training center inventories (A ready reference of what is available on loan)

- Professional magazines (A good medium for project reports, contact persons, new products and innovations)

- Surveys and questionnaires (Personal follow-up proves most effective)

- Field personnel (These are the eyes and ears which bring back what's happening out there)

Information for input into the system can be actively sought or can flow voluntarily from contributors. Several means for collecting information should be evaluated and the most effective ones chosen. Fire service newsletters will often contain information or contact people who can provide material or ideas to be entered into the system.

Cataloguing and Storage

These internal processes direct how the information, once fed into the system, is handled. Obviously, the data must somehow be sorted. Is the resource a printed document, an entire program with props, an audio-visual aid, equipment for loan or a contact person? Then some sorting or cataloguing by topic matter seems advisable. Perhaps a review or evaluation process is included. Is the resource available for storage at a regional or central location or retained by the owner? Sophistication and complexity will be

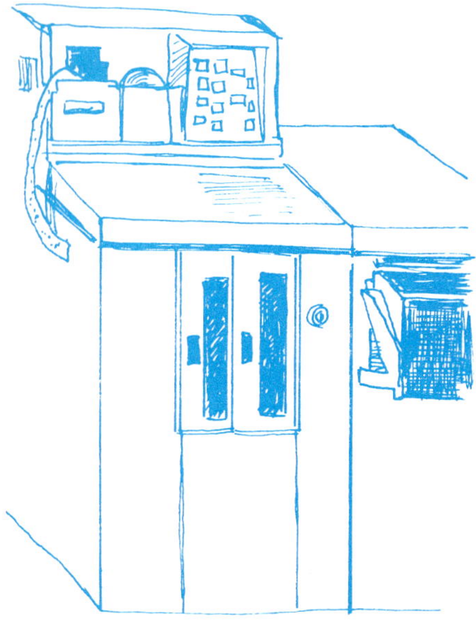

The method chosen to catalogue and store available information will affect how quickly and easily the information can be retrieved. Consider this aspect of the system carefully and choose the storage means which is the most feasible and efficient.

dependent upon funding and personnel. Some effective methods are:

- An index card system (Card or roll file)
- Shelf filing (Label and store alphabetically)
- Log sheets (Ready inventory)
- Keypunch and computer entries
- Local, regional or central storage (Accessibility)
- Microfilm
- Personal knowledge

Topical categories for public fire education programs used in the sorting process might be identified as follows:

Preschool	Elementary school
Junior/middle/high school	Juvenile fire-setters
Home inspection/surveys	Fabric flammability
Smoke detectors	False alarms
Arson	Elderly
Handicapped	Home fire safety
Burn prevention	Outdoor fire safety
Babysitter programs	Urban
Rural	Working with news media
Audio-visual skills	Training fire educators in
Community groups	the community/auxiliary
High-rise	activities
Fire education planning	Fire education evaluation

Retrieval and Dissemination

Retrieval and dissemination procedures hinge largely upon the previous two components. The manner in which data was collected and stored has an effect upon the ease in which data is pulled out of the system. If a card index was used to record, sort and inventory resources by topic, the same index can be used to respond to requests for information. Methods used to gather information and encourage contributions serve doubly as a way to distribute information. If a newsletter is used to keep up with the program development and contact persons, updated inventories and progress reports may in turn stimulate others to submit information for the next issue. Some additional dissemination approaches may be through:

- Conference, workshops and seminars
- Program and equipment displays
- Special resource exchange sessions
- Personal technical assistance

Conferences and seminars are ways to disseminate information. Attendees can gain valuable information and programs which they can use in their public education efforts. Attendees may also have skills or information to share and this, in turn, will benefit the entire Resource Exchange System.

Feedback and Maintenance

This should be an on-going effort of all participants in the system. Updating of information should be readily built into the system due to the circular flow pattern; however, the processor/ manager should take heed of the changing needs of the participants. As an example, perhaps published indexes should be printed more often or the system should expand to include another geographic area.

SYSTEM COMPLEXITY

The degree of complexity of a Resource Exchange System is totally dependent upon the needs and desires of the participants involved. It can be as simple as a sharing session over lunch by a group of public education specialists or as structured as a computer network with local terminals. Some factors which should be considered are:

- The number of resources involved
- The number of information contributors and users participating
- Geographic boundaries or areas served
- Scope of interest groups involved (within fire service and community wide)
- Funding limitations
- Administrative personnel
- Storage restrictions
- Goals, objectives and intent of system

The simplest system is witnessed routinely in fire departments in the form of one-to-one communications between staff members, perhaps over a cup of coffee. Staff meetings are a good way to share experiences and provide an opportunity to preview project completions. Often departments will print an inventory of audio-visual aids and equipment available on loan to others. This promotes reproduction of existent programs with little or no change. Joint usage of slides, films and projection equipment may lead to a designated storage area accessible to all. Regional centers or pooling of resources may result.

Expanding resource exchange outside the perimeters of the fire service population increases the complexity of the system. Because more potential contributors and users are involved, the components of the system must be more flexible to bring together community and fire protection personnel. Distribution of newsletters, inventories and announcements of meetings, etc., become more massive. Coding of bulletin articles, cross-referencing of topics and computer assistance may be desirable. Needless to say, as more resources are gathered and more transactions are completed, the exchange process becomes more complicated and requires a more structured format to be efficient.

System planners may want to ask themselves these questions: "What are the goals of the exchange process? Do we want to reach out and stimulate others to get involved in public education by providing leads to successful accomplishments and available tools, or is the system existent only to collect data should someone make a request? Should the system concern itself only with materials, resource persons or programs? Is the system the only one of its kind in the area or are others available?"

Start out with a few simple features to get a basic system started. See how it operates then gradually tackle a different mode to increase participation. For the most part, the needs will dictate the degree of complexity.

THE PEOPLE INVOLVED IN RESOURCE EXCHANGE

As previously discussed, when resource exchange occurs on a limited one-to-one basis, it involves two parties — the contributor and the receiver (or user). On a larger scale, a Resource Exchange System planned to stimulate more involvement incorporates at least a third party — the information processor/systems manager.

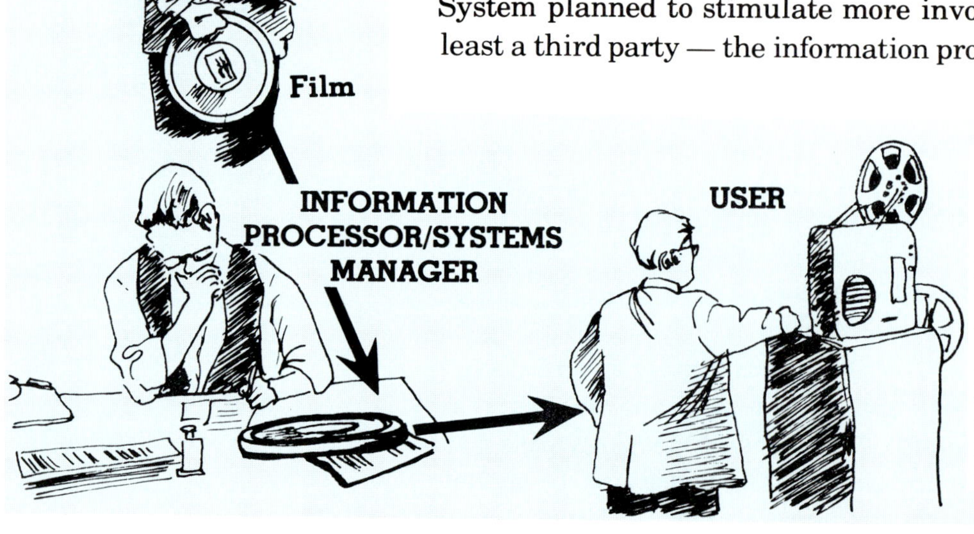

As a Resource Exchange System becomes more complex, an information processor/systems manager may be required. This person receives the input, evaluates, catalogues, and stores and then disseminates the information to the users. One person may handle both the responsibility of information processor and systems manager or, in larger systems, two people may be required to fulfill both jobs.

THE INFORMATION PROCESSOR/SYSTEMS MANAGER

The person who oversees the administration of the Resource Exchange System selects and designs features in line with established goals. He evaluates and measures systems accomplishments. Operational processes are implemented and supervised by the Systems Manager.

The person who physically handles the information is the Information Processor. The Information Processor takes the data contributed, reviews it, sorts it by category, indexes it and makes arrangements for storage as necessary. He then is instrumental in retrieving the information for distribution. The Information Processor operates under administrative guidelines established by the Systems Manager. In most cases, unless the system is extremely large, the Information Processor and Systems Manager are one and the same person. He serves in a facilitating role to enhance resource exchange, seeing that valuable information is passed on to those who can most benefit from it. His primary job is to decrease the constraints of users to increase involvement. He is concerned with how to get action in public education. By eliminating the constraints and giving people access to useful aids, people are encouraged to tackle fire education projects. Getting information is the first step. Application based upon local capabilities follows.

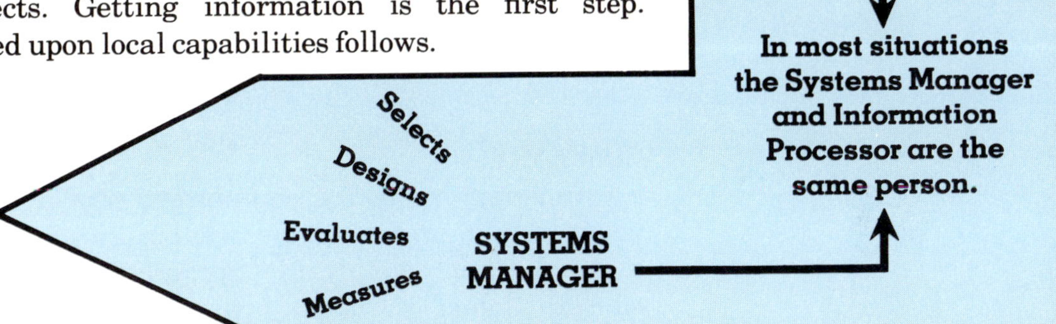

The following planning document may be helpful to Information Processors and Systems Managers:

Information System Management Worksheet
- Who participates in your state communication network?

- How do you involve participants in the network?

- What constraints does your network face?

- Which of these constraints can be overcome? How?

- How do you get information into the network, i.e., how is information collected and stored?

- How can you retrieve and disseminate the information to members of the network?

- How is the information updated?

- How do you achieve feedback?

- What can you do now? In six months? A year from now?

THE CONTRIBUTOR

The contributor is the person who has something of value to offer. He has information, a knowledge, a skill or an experience to share. He may have a program, a slide presentation which can be copied or a projector available for loan. He may have had a bad experience he wants to share so others can avoid the same pitfalls. He may have a written plan of action which worked. These are all resources which can be fed into the system so others may gain from them. The intent is to help others make their job a little easier, maybe less costly, and hopefully more successful in the end.

THE RECEIVER (OR USER)

The receiver may, after application, become a contributor

The receiver is the person who immediately benefits from the exchange. He learns of new methods to reach target audiences. He can call up a contact person and get first-hand assistance. He can copy a recorded radio message without producing his own. Teacher packets may be duplicated with only slight revisions. In turn, the user of the resource takes the information, applies it and feeds back into the system to keep others apprised of his program, his successes or failures. When that cycle is completed, another resource packet may become available and can be entered into the system.

SUMMARY

In summary, let's discuss briefly why resource exchange is needed. As in any relatively new but growing field of expertise, demand exceeds supply. Most programs designed specifically to educate "John Doe Citizen" are still in their infancy. The number and availability of such programs are minimal thus far in comparison to other aspects of fire protection.

Feedback into the system may modify an existing program and increase its success

In an area which is yet being pioneered, public fire educators are desperately searching for some way to attack the base of the problem — citizen apathy regarding the fire problem and ignorance of safe practices in daily environments. Because we are striving to create attitude and behavioral change, few public education planners can feel confident that a specific approach will be successful. Needless to say, very few of us have over-generous budgets to indiscriminately try out different ideas.

Evaluation of programs often is not complete until several years following implementation, but periodic measurements of success experienced by others in similar situations do generate enthusiasm and commitment. After systematically analyzing the problem, resource exchange gives planners an opportunity to more thoroughly research existent programs and learn of outcomes, thereby increasing the probability of success.

Being able to communicate and relate to others with similar goals spurs interest, challenging others to get involved. Together, public educators can stay in step with innovative processes, use newly researched and tested tools, sharing in efforts to impact upon the fire problem in each community.

Blueprints for Planning Material from United States Fire Administration.

BLUEPRINT FOR PLANNING:

	INFANTS TODDLERS	PRESCHOOLERS
AGE	0 1 2	3 4 5
FIRE AND BURN HAZARDS	• Hot liquids, scalds • Hot objects, burns • Space heaters • Flammable clothing • Electrical cords and outlets • Cleaning agents and other poisons within reach • Radiators and floor furnace registers • Open flames	• Playing with matches • Climbing on stoves • Splashed hot liquids • Flammable clothing and costumes • Appliances pulled down and touched • Cleaning agents and other poisons within reach • Hot bath and tap water • Open flames • Smoke
PERSONAL GROWTH	• Emotional problems may set the stage for later fire setting behavior	• Children imitate others' actions including smoking and cooking • Active and curious but still dependent, don't really understand rules about fire use
EDUCATIONAL APPROACH	• Teach parents and babysitters about fire safety for babies. Stress supervision and how to make homes safe for babies • Organize fire safety workshops for professionals who work with parents • Be extra careful that babies are supervised when adults are under stress or preoccupied	• Demonstrate the dangers of fire • Teach children to drop and roll and cover their faces • Teach children the proper use of matches • Pay attention to children imitating adult use of fire • Establish limits but don't rely on rules • Teach children that their clothing is flammable
		SMOKERS WITH CHILDREN BE ESPECIALLY CAREFUL
AGE	0 1 2	3 4 5

FIRE EDUCATION FOR CHILDREN

ELEMENTARY SCHOOL

|6 7 8 9 10 11 12|

- Playing with matches
- Flammable liquids
- Cooking at stoves
- High tension wires
- Flammable clothing and costumes
- Candles
- Campfires and barbeques
- Flammable liquids
- Fireworks
- Mini-bikes, lawn mowers and snowmobiles
- Smoke

- School age children can understand the causes and consequences of fire, are often willing to work hard, earn recognition and improve skills, so discourage fire play and explain fire dangers

- Teach children to call the fire department
- Teach children to look for and correct fire hazards
- Practice emergency home escape drills and learn to use fire safety devices
- Hold home community hazard checks sponsored by the fire department

THIS IS THE AGE WHERE HAZARDOUS FIRE PLAY IS MOST LIKELY

|6 7 8 9 10 11 12|

TEENAGERS

|13 14 15 16 17 18|

- Cigarettes in combination with alcohol and drugs
- Stoves and cooking
- High tension wires
- Explosive chemicals
- Experimentation with gasoline and motor vehicles
- Fireworks
- Mini-bikes, lawn mowers and snowmobiles
- Smoke

- Time to think about careers and the future, so put fire safety in those terms. Remember teenagers have a tendency to hero worship, tend to test authority and be influenced by peer pressure

- Summer job programs to clean up fire hazards in the community
- Home safety education for babysitters
- Include fire education in health, home economics, career planning and parenthood classes
- Don't threaten. Educate about proper procedures, and teach them that they can be of assistance to others

|13 14 15 16 17 18|

SPECIAL NEEDS

- Children with physical and mental handicaps are more prone than others to be victims of fire and burns
- Blindness: Braille, escape plans and fire information; practice exit drills
- Deafness: Bright lights as fire alarm

- Retardation: Bright lines on floors to aid exit drills
- Physical handicaps: Ramps for wheelchairs, special doors and windows
- Children from low income families often face greater hazards in the environment
- Special instructions often may be needed for children from non-English speaking families

- Parents are very receptive to home safety information. During pregnancy, they can be reached in obstetricians' and clinic waiting rooms. Parents tend to feel more responsible for a first baby and are more open to information, counseling and advice

BLUEPRINT FOR PLANNING: FIRE EDUCATION FOR ADULTS

ADULTS ARE THE KEY

TO FIRE PREVENTION

- They start more fires than any other age group
- They put out more fires than any other age group
- They are the relatives of younger and older people who are burned
- They are burned seriously least often

and
THEY ONLY HEAR
WHAT THEY WANT TO.

They are:
- Parents • Workers • Employers
- Tenants • Teachers • Firefighters

They are the people who are supposed to be responsible.

ADULTS THINK: IT WON'T HAPPEN TO ME . . .

THE TRUTH IS: IT MIGHT . . .

HOW CAN YOU CHANGE THAT ATTITUDE?

(1)

TELL THE FACTS

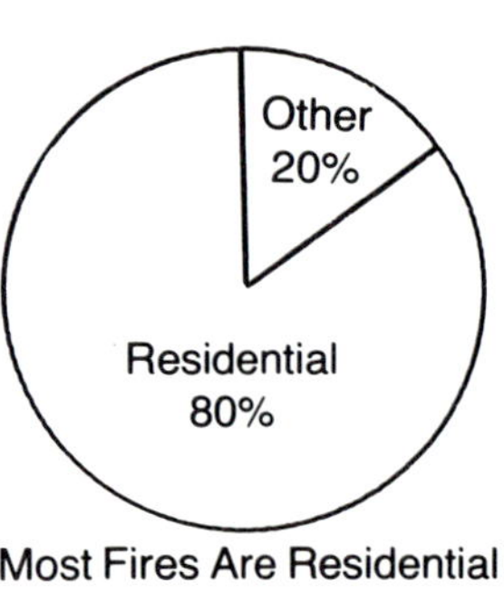

SMOKING IS A MAJOR CAUSE OF FIRES AND BURNS IN THE UNITED STATES

12,000 LIVES ARE LOST BY FIRE AND BURN INJURY EACH YEAR

3 OUT OF 4 HOUSEHOLD FIRES ARE DISCOVERED BY WOMEN!

A PERSON IN THE U.S. IS LIKELY TO BE INVOLVED IN THREE FIRES IN A LIFETIME WHICH WILL REQUIRE THE FIRE DEPARTMENT

3 OUT OF 5 FIRES INVOLVE APPLIANCES. MORE THAN HALF OF THESE INVOLVE THE IGNITION OF GREASE OR OTHER FOODS

9 OUT OF 10 RESIDENTIAL FIRES ARE PUT OUT BY A MEMBER OF THE HOUSEHOLD, USUALLY A WOMAN

TODAY 33 PERSONS WILL LOSE THEIR LIVES BECAUSE OF FIRE

(2)

APPEAL TO BASIC CONCERNS
BY ENVIRONMENT

RURAL

"I have to be my own fire department."

SUBURBAN

"Has everyone practiced our home escape plan?"

URBAN HOMEOWNER

"Is our neighborhood free of trash and debris?"

URBAN TENANT

"Does my building have fire escapes?"

HIGH RISE

"How can I get out in case of fire?"

(3)

TEACH FIRE SAFETY

- Common fire hazards
- What to teach children about fire
- What to check before moving into an apartment or house
- How to make home safe for babies
- What to tell babysitters about fire
- How stress, fatigue and intoxication increase fire risk
- Emergency fire and burn procedures
- How to respond to kids who set fires
- Home maintenance repairs

ATTENTION!!

People with physical, emotional or mental handicaps are more prone to be injured or burned and require special attention.

However you approach fire education, follow your own sense of what works best in your community.

In low income areas, extra effort is needed to keep the community fire safe.

There are some times in peoples' lives when they are especially receptive to new information and lifestyle changes. One such period is just before the birth of a baby. Another is after a local fire tragedy.

BLUEPRINT FOR PLANNING: FIRE EDUCATION FOR OLDER AMERICANS

WE ARE 20 MILLION AMERICANS OVER 65
70 75 80 85 90 95 100

WHO ARE WE?

WE ARE WHO WE USED TO BE
ONLY OLDER AND PROBABLY POORER

- 67% of us own our homes, but many are in substandard conditions

- Only 5% are in nursing homes at any one time

- 86% have chronic health problems and average 14 days/year in the hospital

- 60% live in metropolitan areas

- More than 1 in 4 live on less than $1,850 annually

- 50% were born in another country

- 90% are registered voters and 67% vote regularly

> ### DON'T CALL ME OLD!
> Like many others, we are often prejudiced against aging. Many of us don't like to be called "old," "elderly," or "senior citizens." Refer to us as you would anyone else.

WHAT DO YOU NEED TO KNOW ABOUT US?

- We can remember things that are meaningful

- Tell us, demonstrate and let us experience what you are saying

- Many of us believe we can't afford fire safety devices

- We may resist home inspections because we are afraid of letting strangers into our homes

- We must balance our need for security with the ability to escape from a fire

- Careless smoking, combined with alcohol and illness or fatigue, is our No. 1 cause of fires and burns

- Many fires are started when we suffer minor strokes, heart attacks or because of diminished sense of sight, hearing, touch or smell

- Most of our burn accidents occur between 7 a.m. and 10 a.m.

- We tend to wear loose fitting clothing which is especially flammable

WHAT DO WE NEED TO KNOW ABOUT FIRE

FIRE AND BURN HAZARDS

SAFE USE OF APPLIANCES:

- Don't store food in ovens; do use cabinets
- Turn off gas stoves between 1st and 2nd matches
- Don't dry clothes on heaters or radiators; do use towel bars and clothes racks
- Don't wear loose fitting clothing around stoves; do wear robes with tight fitting sleeves
- Don't leave unattended pots on stoves; do stay in kitchen when cooking

REMEDIES FOR SUCH PROBLEMS AS:

- Making our homes safe for grandchildren
- Appropriate agencies to contact when a landlord can't or won't keep building up to code
- Emergency home escape planning
- How to get help to eliminate fire hazards

HOW CAN YOU INTEREST US IN FIRE SAFETY?

APPEAL TO OUR:

- Desire to remain independent
- Desire to protect possessions and health
- Desire to have a safe environment for visiting grandchildren and other relatives
- Desire to better guide and advise younger people
- Desire to be a better informed citizen
- Increased sense of responsibility

HAVE FUN

When you organize an "event", bill it as a festive occasion with food, fun and fire education

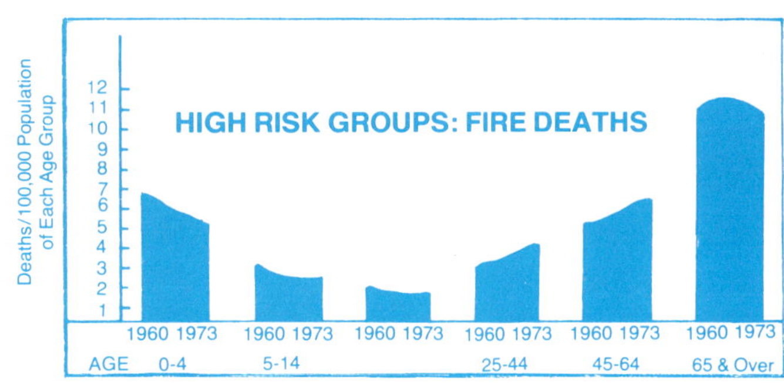

SAMPLE PROGRAM IDEAS

Educate people who already have contact with us: local home care workers, social workers, medical professionals and others interested in our welfare.

Organize retired graphic artists, crafts people and others to make fire education posters. Help place posters where we spend time.

Create a community cooperative to buy smoke detectors and other safety devices at a discount. Use this program as a basis for home hazard checks while going door to door.

Organize mutual aid groups to deal with fire problems. In high rises, assign captains on each floor to explain escape plans and every fire safety feature of our building.

BLUEPRINT: FIRE EDUCATION AND

AGE	1	2	3	4	5	6	7

LEARNING ABOUT FIRE
IS A PART OF EVERY CHILD'S GROWTH PROCESS

Children begin experimenting with lighting matches.

Maximum frequency of match and fireplay. Guilt, fear in response to big fires.

DANGEROUS FIRE SETTING BEHAVIOR
CAN BE A SIGN OF UNHAPPINESS AND ANXIETY

Onset of problems which may result in fire setting in some deeply disturbed children.

Rage reaction to troubles at home. Some children absorb parents' anger and express it with fire.

AGE	1	2	3	4	5	6	7

WHO ARE THE YOUTHFUL FIRE SETTERS?

- Children, usually boys, who are unhappy and have poor self images and feel lousy about their situations

- Children who have troubled home lives, conflict with parents and brothers and sisters

- Children who do not function well in school and have conflicts with teachers and peers

- Children who have no means of expressing their anger except through their behavior

- Children whose verbalization skills are not sufficient to express their unhappiness

COMMON CAUSES OF FIRESETTING
- Boredom
- Vandalism
- Peer pressure
- Curiosity

FIRE SETTING IS A CALL FOR HELP. IF NOT HEEDED THE SITUATION WILL ONLY GET WORSE

YOUTHFUL FIRE SETTING BEHAVIOR

8   9   10   11   12   13   14   15   16   17   18

Fairly normal, disobedient fire play.

Risk-taking actions based on feelings of immortality or lack of knowledge.

Rage response to troubles in school. Lack of support and affection at home seen in more serious fires which sometimes cause great damage.

Identification of father with firefighters. Pleasuring in helping to put out fire.

More knowledgeable and daring firesetting causing death and destruction.

8   9   10   11   12   13   14   15   16   17   18

HOW TO SPOT A POTENTIAL FIRE SETTER:

LOOK FOR THOSE CHILDREN WHO EXHIBIT ONE OR MORE OF THE FOLLOWING CHARACTERISTICS:

- A child who has great trouble verbalizing feelings, who may steal, be cruel to animals and/or seem unusually self-destructive.
- A child with severe emotional or physical handicap
- A child whose behavior suddenly changes

- A child with poor judgment and poor impulse control who may be accident prone or hyperactive
- A child who has a school behavior problem, has a hard time making friends, may be a bully and is usually a loner

PREVENTION

- Be willing to talk with parents who call seeking help
- Help children change their self image so they won't feel a need to set fires
- Educate about fire safety to prevent "thoughtless" fire setting
- Suggest professional counseling to help children work out problems which motivate firesetting

INTERVENTION

- Trip to the firehouse with school class, scout troop, etc., to meet firefighters
- Fire education to improve knowledge and judgment
- Individual counseling to find out what is troubling child
- Trip to burnt out building where someone was hurt
- Family counseling or referral to appropriate service agency to determine the extent of family or individual treatment required
- Referral to psychiatric and legal agencies
- Counseling for other family members

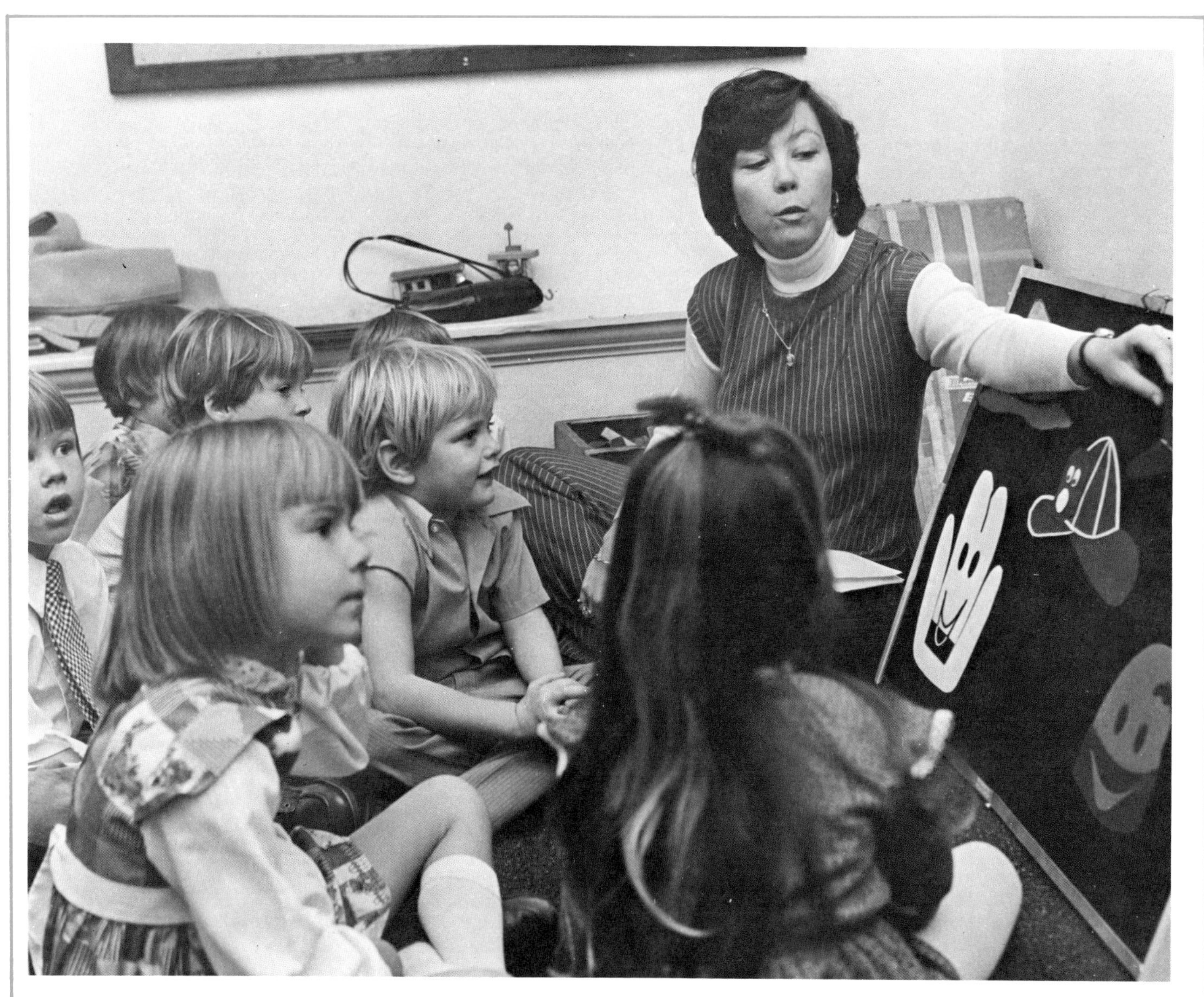

Photograph courtesy of Oklahoma State University *Outreach* Magazine

Appendix B
BURN INJURIES

Each year in the United States approximately two million people, or 7.5 persons per 1,000, sustain burn injuries severe enough to be seen by a physician. Two hundred thousand of these individuals are admitted to a hospital and perhaps twelve thousand will die as a direct result of a burn injury. It is important to note that 87% of all fire deaths occur in the home; 20% of those occur between the hours of 11 p.m. and 7 a.m. while people are sleeping. Fire takes more children's lives than do childhood diseases. Forty-five percent of all fire deaths occur in children under five years of age, while 20% occur in the elderly. Any size burn occurring to those people in the extremes of age (either less than one year or greater than 65 years) or to persons having underlying disease processes, can be associated with significant mortality. Mortality for all ages depends upon age, depth of injury, total percent of the body surface burned and pre-existent disease.

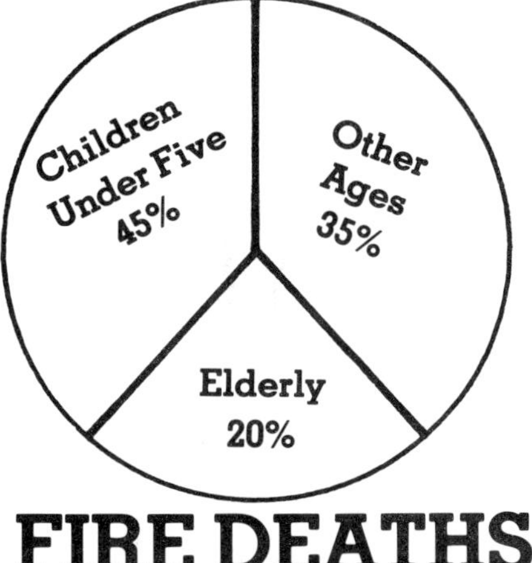

FIRE DEATHS

Facilities for caring for the burn patient range from general hospital beds, under the care of a physician who has an established plan of care for the burn patient (burn service), to highly specialized units or larger burn centers where many severely burned patients are treated each year. In the United States there are 1,509 beds designated strictly for the care of burn patients. Although some hospitals are able to treat burn patients skillfully without specially allocated burn beds, it is generally agreed that facilities for treating these severely injured individuals are not yet adequate in number or size. The cost of intensive care for the burn patient in 1978 ranges from $325 to $1825 per day, and severely injured patients may need this type of care for several months. Many types of specialized supplies, equipment and services are necessary to provide comprehensive care for these patients, and hospital bills can rapidly reach astronomical figures. The end of hospitalization does not mark the end of care for the burn patient; follow-up care and rehabilitative services are often needed for years after the injury. In many cases, reconstructive procedures are necessary to improve both cosmetic results and function of an injured part. Thus, the road to recovery for the burn patient is a long, complicated and expensive one. It is evident that there is no substitute for prevention of fire and for taking all measures to minimize injuries and complications that may result.

TYPES OF BURN INJURIES

THERMAL

Burn injury can be caused by a variety of destructive forces and vary somewhat depending upon the causative agent.

Thermal burns are caused by flame, steam or hot liquids. In the home, cooking accidents are common and result in scald burns that often involve children. The severity of the injury produced by a thermal burn depends on the temperature of the burning agent and the length of time it is in contact with the skin. Reduction of either of these factors will reduce the severity of injury. Causes of thermal injuries often result from faulty heating systems and open fires or fireplaces. In warm weather, outdoor activity devices are often fire hazards, such as barbeque grills or gasoline-powered engines in boats or lawn mowers.

A particularly dangerous complication of thermal burns is injury to the respiratory system or airway, known as an inhalation injury. The destructive agents causing these injuries are usually products of combustion rather than direct flame. Breathing smoke, fumes or even very hot air can cause injury to the airway, resulting in swelling and extreme difficulty in breathing. Inhalation injury is more likely to occur if the fire is in an enclosed space, such as a house fire, because under these circumstances a higher concentration of noxious gases and smoke is present in the air. Inhalation injury can be very serious, and the patient may require respirator support in order to assist breathing until swelling is reduced and he is able to breathe independently.

ELECTRICAL

Electrical burns are caused by the heat produced by passage of current through the body. Typically, an entrance and an exit wound are visible, with variable amounts of tissue destruction at these two sites. Initially, it is often impossible to know the extent of injury because the overlying skin may be normal and conceal dead tissue. Only after careful clinical observation of the patient and diagnostic workup is the extent of the damage apparent. In some cases, especially those involving injury to the extremities, surgical exploration may be necessary to determine the extent of tissue damage. The severity of an electrical injury depends upon the surface area of the point of contact with the current, the length of time of contact, and the amperage of the current.

CHEMICAL

Chemical burns are caused by skin contact with caustic chemicals. Household chemicals, both alkali and acid, that may cause injury include lye preparations, drain openers, battery acid and other corrosive substances. Industrial accidents may also result in chemical burns. Length of exposure to the substance is the major factor determining the severity of a chemical burn. The pattern of the burn may be spotty or it may cover a large surface, depending on how the agent came in contact with the skin. Tissue damage is often deep, especially if the substance was not immediately and completely removed. A particular hazard of chemical burns is that of swallowing some of the burning agent, resulting in death or extensive damage to the gastrointestinal system.

Radiation burns are caused by overexposure to X rays, radioactive substance, nuclear blast or sunlight. The most commonly seen radiation burn is a sunburn. Usually sunburns are very minor injuries, although if blistering occurs the injury is more serious and should be treated.

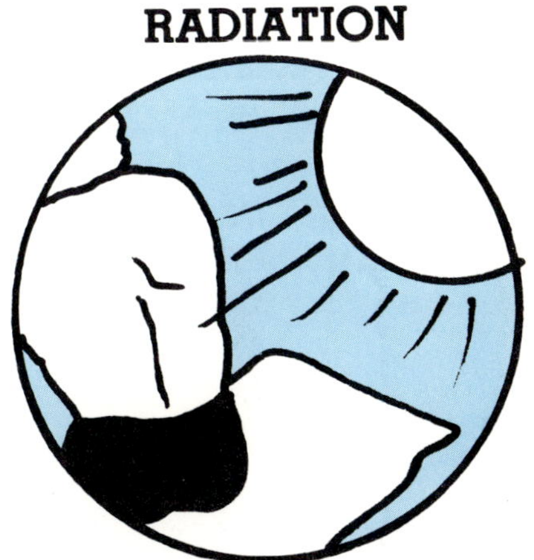

ASSISTING THE BURN PATIENT

Very few burns are so minor that they do not require some medical evaluation and treatment. Any initial burn of the face, feet, hands, eyes or genital area should receive professional treatment. Whenever any doubt exists regarding the severity or proper treatment of an injury, medical treatment should be sought.

The first step in dealing with a burn injury is to stop the burning process, including extinguishing flaming clothing and removing heat-retaining objects such as belt buckles. When possible, it is advisable to cool the burned surface with water in order to reduce further tissue destruction in the heated tissues. Ice should not be used to cool the burned area because it may cause further damage to already injured tissues. Covering the burned surface with a clean sheet is all the local treatment that is necessary or desirable. The cover over the wound significantly reduces pain by preventing air contact, and it protects the wound from dirt and bacterial contamination. Medical evaluation is best performed on a clean wound free of any ointments or other applications. It is also advisable that blisters be left intact because they provide an excellent protective covering until the wound can be properly evaluated.

In the event of a chemical burn, it is imperative that the caustic substance be removed as quickly and as thoroughly as possible. Generous amounts of water should be used in an attempt to dilute and rinse the substance from the skin. Should the chemical be swallowed, the mouth should be thoroughly rinsed. Chemical contact with eyes requires vigorous flushing with large quantities of water. Specific neutralizing agents may be determined by using a poison control chart, the instructions on the substance container, or by calling the nearest emergency facility.

When assisting the victim of a relatively major injury involving the surface area equivalent to an entire arm or leg (half that in children), the basic principles of emergency care for any traumatic accident would apply. One begins with evaluating the ABC's (airway, breathing, circulation) and supporting them when necessary. The ABC's are particularly important for patients involved in house and other closed-space fires where the chance of an airway injury is greatest. Burns of the face, singed nasal hair, and soot in the mouth or sputum signal a high likelihood of airway injury. Airway adequacy and breathing should be evaluated in these victims frequently, and basic cardio-pulmonary resuscitation (CPR) provided when necessary. The burn patient should not be allowed to eat or drink.

Electrical burns may also require special attention. First, the patient should be removed from the source of electricity if contact is still maintained. The rescuer should never touch the individual who is still in contact with the electrical source, but should break contact using any object which is a non-conductor. The ABC's need careful monitoring in these people too, because the extent of internal damage is unknown and electrical burns may be particularly prone to heart complications requiring emergency support. Water should never be used on the electrical injury because the area in which the injury occurred may be extremely hazardous.

Any details of the incident that can be provided may prove helpful to the health team personnel in defining the plan of treatment.

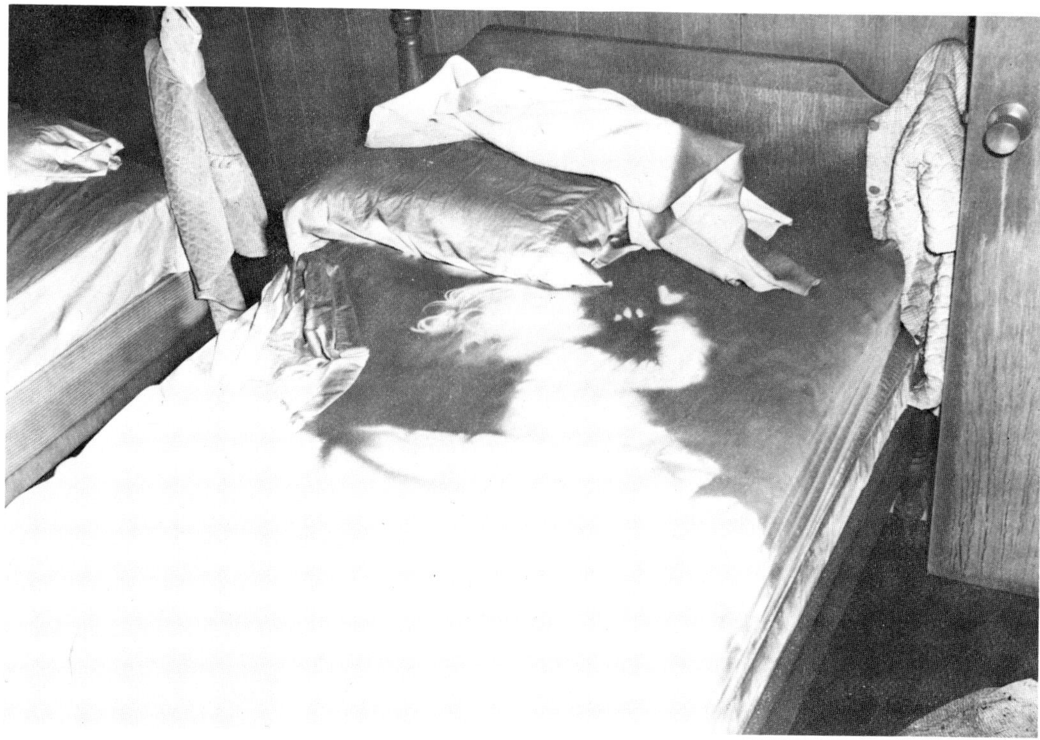

Fatal burn injuries are tragic. Severe burn injuries that scar a child for life can be almost as tragic.

HOSPITAL TREATMENT

Once the burn patient reaches the hospital, definitive treatment begins. Intravenous catheters are inserted and IV solutions administered. During the first 24 to 48 hours, burn patients require large volumes of fluid to replace fluid that leaks from the injured blood vessels into the tissues. This fluid shift causes tremendous swelling and may interfere with breathing or with blood supply to the extremities. A urinary catheter is inserted to measure urine output, which is a good indicator of whether or not the fluid replacement is adequate. The patient is carefully examined and blood gas tests performed to determine if oxygen exchange is adequate and whether or not airway injury has occurred. In some cases oxygen and/or mechanical breathing equipment may be required. Pulses are checked frequently because increased swelling may reduce blood supply to burned extremities. Surgical incisions may be required to release extreme pressure on the blood vessels and insure adequate circulation. Any other injuries resulting from the accident are evaluated.

The first step in wound care involves cleaning with a mild soap, carefully removing blisters and carefully evaluating the depth and the total percentage of the body surface burned. An accurate figure for total body surface area burned is necessary in order to treat the patient appropriately with intravenous fluids. An estimate of the total percentage of the body burned can be made quickly using the rule of nines. According to this method, each part of the body is viewed as a multiple of nine percent of the total. Each leg is equivalent to 18%; each arm, 9%; front and back body, 18% each; and head, 9%. One percent remains for the genital area. A more detailed adult and child chart is used for precise evaluation of all burns, and is necessary because adult and child proportions are quite different.

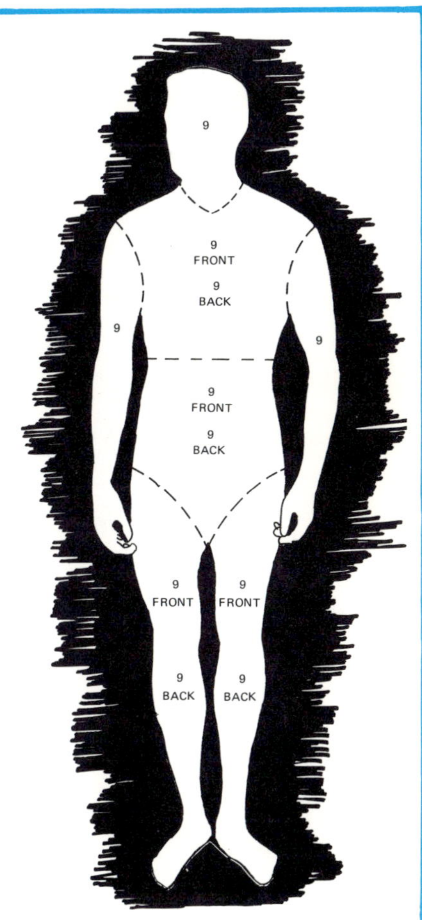

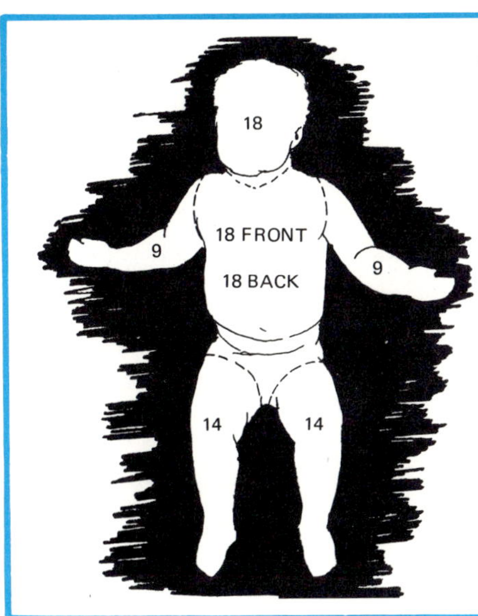

Using the rule of nines, a quick estimate of the total percentage of the body burned can be made. Note the differences in percentages between an adult and a child.

Depth of injury can be difficult to determine initially. Burns may be generally classified as partial-thickness or full-thickness injuries. Partial-thickness burns may involve only the outer layer of skin (first degree) or may involve the outer layer and part of the dermis (second degree). Second degree burns vary widely with regard to the depth of tissue destruction. Some deep second degree injuries may require skin grafts for best results. Full-thickness or third degree burns destroy both the epidermis and the dermis and will not heal without skin grafts.

The patient's wounds are cleaned thoroughly, usually in some type of hydrotherapy tank or tub. Loose, dead tissue is removed, or debrided. Initially, topical chemotherapeutic agents are applied to reduce the chances of a serious burn wound infection. After a period of topical therapy, other forms of wound dressings are used. Wet dressings may be used to remove bits of dead tissue, or eschar. Protection for the healing wound may be provided by biologic dressings including pigskin, human donor skin and amnionic membrane. Some synthetic dressings are also used. All dressing procedures are aimed at promoting more rapid healing, or in the case of full-thickness injury, faster preparation of a tissue bed for skin grafting. The realm of burn wound care is constantly changing as researchers continue to attempt to develop improved methods of caring for the wound.

Great precautions must be taken to prevent infection in a burn injury

In cases of full-thickness burn injury, skin grafting is always necessary. There are no skin cells left to regenerate, and healing would otherwise have to occur from the wound edges, an extremely slow and painful process. Permanent skin grafting requires a surgical operation in which strips of skin are removed from the patient's unburned areas and placed on the clean open wounds. These strips will eventually grow and form permanent skin coverage. Frequently skin strips from patients with large injuries will be "meshed" or have a series of slits cut in them similar in appearance to a honeycomb. These meshed grafts may be stretched over a larger surface. The grafted skin then grows and fills in the holes. The area from which the skin was removed, or donor site, usually heals in 12 to 14 days. Occasionally, these donor sites are used again when there is a large surface to be grafted. For a skin graft to grow successfully, or "take", it is necessary that the wound be healthy, free of dead tissue and have minimal bacterial growth. In some cases, human cadaver skin may be used as a test graft to determine whether or not the wound is ready to accept a permanent skin graft.

Skin grafting is always necessary in full-thickness burn injuries

One specific method for preparing a full-thickness wound for grafting involves surgically removing the dead tissue early in the hospital course. The resulting wound is then covered with a dressing (usually human cadaver skin) or grafted immediately. If

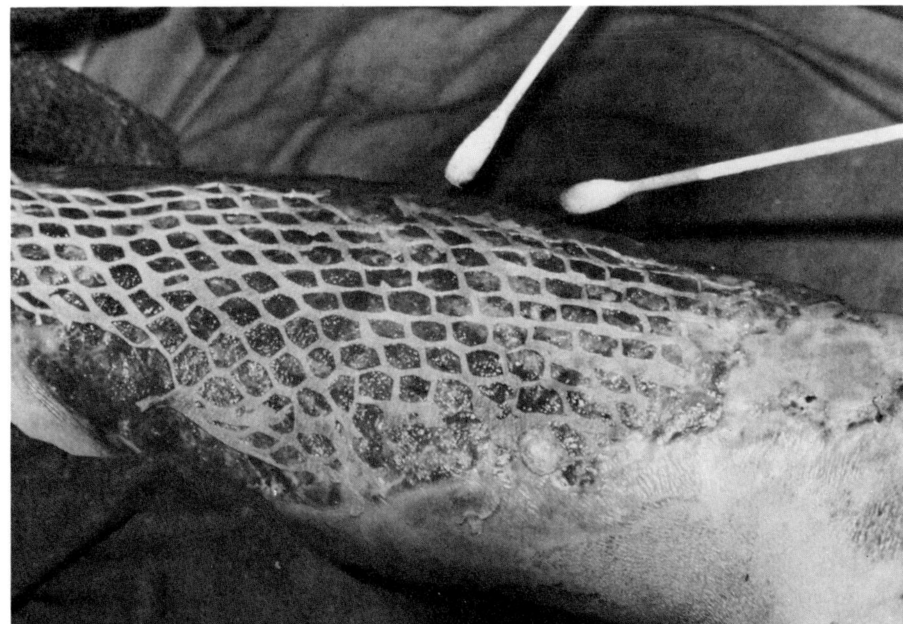

Meshed skin grafts can be stretched over larger surfaces and are often used in large injuries. The grafted skin will eventually grow and fill in the holes.

a dressing is used, the wound is examined for readiness to accept a graft and permanent skin grafting is carried out as soon as possible. This method is increasing the survival rate of many patients with large, full-thickness burns.

The course of recovery from a burn injury may often involve special problems or complications. The greatest hazard that the burn patient faces is that of infection. Meticulous wound care, judicious use of antibacterial medications, and strict isolation procedures are necessary to prevent devastating infections. Lung complications can often occur, especially in patients who have sustained an airway injury. Immobility of the patient is often required during periods of burn treatment and increases the likelihood of pulmonary complications such as pneumonia, and contributes to other complications including pressure sores, urinary tract infections or additional wound degeneration. The extreme stress of the injury can lead to gastrointestinal erosions, known as Curling's ulcers. Burns place extreme nutritional requirements on the patient in an attempt to compensate for heat and water loss and to repair tissues. Significant weight loss over the course of hospitalization is not uncommon.

Maintaining function of the burn parts requires special attention during the hospital, and long after the patient returns home. Intensive physical therapy is essential and requires a great deal of independent work on the part of the patient. Special devices and splints are often used to position burned extremities so that maximum functional use can be preserved. Many times the proper position is an uncomfortable one and is an added stress to the patient. Occupational therapists provide activities to help patients regain use of their extremities and devise splints, etc. to assist in maintaining function.

An important area of concern in the care of the burn patient is the effect of the psychological trauma of such a devastating injury. Patients may be concerned immediately with survival. Soon thereafter, they must attempt to deal with their appearance and what they will eventually look like. They are concerned with their family, and the stability of existing relationships is sometimes questioned. Burn care facilities are usually very confined environments and visiting is strictly limited. The patient may be hospitalized for long periods of time and suffers from lack of stimulation and loss of contact with the outside world. There is always the question, "Will the patient be able to return to his former job?" Many burn care procedures are quite painful and the stress of dealing constantly with severe pain can be destructive emotionally. Psychological consultants, as well as chaplains, nurses, physicians and any persons who may work closely with burn patients, must carefully consider the patient's emotional needs.

BURN INJURIES CAN SCAR Emotionally As Well As Physically

AFTER DISCHARGE

Once the burn patient returns home, a very difficult challenge is faced. He must now attempt to adjust to a world in which he is different. Disfigurement is far more obvious in a world of normal individuals than it is in the company of other burn patients. Often there is long and difficult rehabilitation before many activities which were once considered routine tasks can be resumed.

Medical care must continue for months and even years after the injury. Scars often form and frequently must be treated with tight elastic stockings and special garments. In areas where there is any movement, burned skin may shorten or contract, causing the part to assume unattractive positions and severely reducing functional mobility. Individuals may be required to wear rigid splints and devices to encourage stretching and proper positioning in an attempt to prevent these deformities. Many times, repeated surgical procedures may be neccessary to reduce scarring and contractures, aimed at improving both the function and the cosmetic appearance of burned areas. Activity limitations may be imposed. Often burned individuals must stay out of the sun, and have poor tolerance to many clothing materials. Heat intolerance is common. Functional limitations may impose undesirable restrictions.

Emotional difficulties may continue at home. The person's appearance continues to change, and often burned individuals are uncertain as to what they can expect during the process of healing and scar maturation. Fatigue and depression are common, and functional deficits often prevent the person's usual work from being resumed. Frequently a new job must be sought or working hours altered. There may be many periods of frustration and overwhelming feelings of worthlessness to deal with.

In short, a severe burn injury can alter one's life dramatically and the effects may range from minor to completely devastating. This appendix has provided an overview of the causes and effects of burn injury. It is by no means comprehensive, but provides a knowledge base for what burns can involve. It is hoped that this knowledge might help to prevent, to minimize, or to assist in dealing with the consequences of such an injury.

IFSTA MATERIALS

101 FORCIBLE ENTRY, ROPE AND PORTABLE EXTINGUISHER PRACTICES
Types of forcible entry tools and general building construction; use of tools in opening doors, windows, roofs, floors, walls, partitions and ceilings; types, uses, and care of ropes, knots and portable fire extinguishers.

102 FIRE SERVICE GROUND LADDER PRACTICES
Various terms applied to ladders; types, construction, maintenance, and testing of fire service ground ladders; detailed information on handling ground ladders and special tasks related to them.

103 FIRE HOSE PRACTICES
Construction, care, and testing of hose and various fire hose accessories; preparation and manipulation of hose for rolls, folds, connections, carries, drags, and special operations; loads and layouts for fire hose.

104 SALVAGE AND OVERHAUL PRACTICES
Planning and preparing for salvage operations, care and preparation of equipment, methods of spreading and folding salvage covers, most effective way to handle water runoff, value of proper overhaul and equipment needed, determining cause of fire, and recognizing and preserving arson evidence.

105 FIRE STREAM PRACTICES
Characteristics, requirements and principles of fire streams; developing, computing, and applying various types of streams to operational situations; formulas for application of hydraulics; actions and reactions created by applying streams under different circumstances.

106 FIRE APPARATUS PRACTICES
Various types of fire apparatus classified by functions; driving and operating apparatus including pumpers, aerial ladders, and elevating platforms; maintenance and testing of apparatus.

107 FIRE VENTILATION PRACTICES
Objectives and advantages of ventilation; requirements for burning, flammable liquid characteristics and products of combustion; phases of burning, backdrafts, and the transmission of heat; construction features to be considered; the ventilation process including evaluating and size-up is discussed in length.

108 FIRE SERVICE RESCUE AND PROTECTIVE BREATHING PRACTICES
Respiratory tract and respiratory hazards; protective breathing equipment and how to use, care, test, and inspect; other protection equipment, including clothing, ambulances, and rescue equipment; rescue methods, techniques, procedures and situations.

109 FIRE SERVICE FIRST AID PRACTICES
Completely revised. Brief explanations of the nervous, skeletal, muscular, abdominal, digestive, and genitourinary systems; injuries and treatment relating to each system; bleeding control and bandaging; artificial respiration, external cardiac compression and cardiopulmonary resuscitation; shock, poisoning, and emergencies caused by heat and cold; fractures, sprains, and dislocations; emergency childbirth; short-distance transfer of patients and ambulances; conducting a primary and secondary survey.

110 FIRE PREVENTION AND INSPECTION PRACTICES
Fire prevention bureau and inspecting agencies; fire hazards and causes; prevention and inspection techniques; building construction, occupancy and fire load; special-purpose inspections; inspection forms and checklists along with reference sources; maps and symbols; records and reports.

200 ESSENTIALS OF FIRE FIGHTING
This new manual was prepared to meet the objectives set forth in levels I and II of NFPA, *Fire Fighter Professional Qualifications, 1974.* Included in the manual are the basics of: hose, ladders, fire streams, forcible entry, rescue, salvage and overhaul, ventilation, fire behavior and science, ropes and knots, extinguishers, protective breathing, fire prevention, fire cause identification, ground cover, communications, water supplies and sprinkler systems.

200 SELF-INSTRUCTION FOR IFSTA 200
SI Self-instruction book for IFSTA 200 *Essentials of Fire Fighting.* Covers the most important points of IFSTA 200, including NFPA 1001 Firefighter I and II requirements. Perfect supplement for formal training.

201 FIRE SERVICE PRACTICES FOR VOLUNTEER FIRE DEPARTMENTS
A general, cursory introduction to material covered in detail in manuals 101 - 110, with the exception of first aid.

202 FIRE SERVICE ORIENTATION & INDOCTRINATION
History, traditions, and organization of the fire service; operation of the fire department and responsibilities and duties of firefighters; fire department companies and their functions; glossary of fire service terms.

203 FIRE SERVICE TRAINING PROGRAMS
Qualifications and duties of the training officer in planning, organizing, directing, and controlling the training program; examples of training centers, facilities, and programs; training curricula — a master outline of hundreds of address sources.

204 PHOTOGRAPHY FOR THE FIRE SERVICE
Camera components and operations, films — their advantages and differences, fundamental principles in taking a good picture, processing the negative and print, controlling light for optimum results, accountability and storage of photographic materials; fire scene photography, using photographs as training aids, enhancing fire prevention and public relations through photography, how the investigator can use photographs, glossary of photographic terms.

205 WATER SUPPLIES FOR FIRE PROTECTION
Importance, basic components, adequacy, reliability, and carrying capacity of water systems; specifications, installation, maintenance and distribution of fire hydrants; flow requirements, flow tests and control valves; sprinkler and standpipe systems.

206 AIRCRAFT FIRE PROTECTION AND RESCUE PROCEDURES
Aircraft types, engines, and systems, conventional and specialized fire fighting apparatus, tools, clothing, extinguishing agents, dangerous materials, communications, pre-fire planning, and airfield operations.

207 GROUND COVER FIRE FIGHTING PRACTICES
Ground cover fire apparatus, equipment, extinguishing agents, and fireground safety; organization and planning for ground cover fire; authority, jurisdiction, and mutual aid, techniques and procedures used for combating ground cover fire.

208 RECORDS AND REPORTS FOR THE FIRE SERVICE
Procedures for compiling information for records, methods of writing and presenting reports, various sample forms for all activities of the fire service.

209 FIREFIGHTER SAFETY
Basic concepts and philosophy of accident prevention; essentials of a safety program and training for safety; station house facility safety; hazards enroute and at the emergency scene; personal protective equipment; special hazards, including chemicals, electricity, and radioactive materials; inspection safety; health considerations.

210 PRIVATE FIRE PROTECTION & DETECTION
Automatic sprinkler systems, special extinguishing systems, standpipes, detection and alarm systems. Includes how to test sprinkler systems for the firefighter to meet NFPA 1001.

301 THE FIRE DEPARTMENT OFFICER
The officer's functions analyzed; decisions, planning, activating, problem solving, delegating authority and supervision; fire fighting activities of the fire officer.

302 FIRE DEPARTMENT — FACILITIES, PLANNING AND PROCEDURES
Fire protection facilities and water supplies; pre-fire planning and fire fighting procedures; dispatching companies, size-up, attack, confinement, extinguishment and overhaul.

303 FIRE SERVICE INSTRUCTOR TRAINING
Characteristics of good instructor; determining training requirements and what to teach; types, principles, and procedures of teaching and learning; training aids and devices; conference leadership.

304 FIRE PROBLEMS IN HIGH-RISE BUILDINGS
Locating, confining, and extinguishing fires; heat, smoke, fire gases, and life hazards; exposures, water supplies and communications; pre-fire planning, ventilation, salvage and overhaul; smokeproof stairways and problems of building design and maintenance; tactical checklist.

401 FUNDAMENTAL PRINCIPLES OF MATHEMATICS
Mathematics oriented to the fire service; fractions, decimals, measurements and weights, percentages and graphs, ratio, proportion, powers, and roots, algebraic and geometric fundamentals and problems.

402 FUNDAMENTAL PRINCIPLES OF SCIENCE
A basic investigation of matter, motion, force, machines, liquids, gases; principles of chemistry, combustion and heat, magnetism, electricity, atomic energy and radiation.

403 LEADERSHIP IN THE FIRE SERVICE
A series of lectures by Robert F. Hamm covering management, supervision, personnel administration, problem solving, and analysis.

500 FIREFIGHTER STUDY GUIDE
Correlates IFSTA publications with NFPA Standard 1001 to guide personnel through the requirements for national firefighter certification at levels I, II, III. Also provides a personal logbook for recording test dates, scores and instructor initials.

502 FIRE SERVICE REFERENCE GUIDE
Each IFSTA manual, except 201 and 403, is individually subject indexed and cross-indexed by general subject; articles from *Fire Command, Fire Engineering, Fire Chief, Fire Journal, W.N.Y.F.* and *Fire News* indexed by subject 1970-76; index of FDIC Proceedings for six years. Separate 1978 *Addendum* indexes an additional year of reference material, plus subject index of IFSTA 101 and 206.

606 PUBLIC FIRE EDUCATION
A valuable contribution to your community's fire safety. Includes public fire education planning, target audiences, seasonal fire problems, smoke detectors, working with the media, burn injuries, and resource exchange.

FIRE PROTECTION ADMINISTRATION
A reprint of the Illinois Department of Commerce and Community Affairs publication. A manual for trustees, municipal officials, and fire chiefs of fire districts and small communities. Subjects covered include officials' duties and responsibilities, organization and management, personnel management and training, budgeting and finance, annexation and disconnection.

INSTRUCTOR GUIDE SETS

Available for manuals 101 - 110, 200, 206, and for the slide program *Fire Department Support of Automatic Sprinkler Systems*. Basic lesson plan, tips for instructor, references. 200 has NFPA Standard 1001 references and pertinent proficiency tests.

SLIDES

Ladder   *Salvage   Sprinkler

2-inch by 2-inch slides that can be used in any 35 mm slide projector; supplements to respective manuals and sprinkler guide sets.

*The complete package consists of the slides, instructor's manual, and instructor's guide sets.

**Smoke Detectors Can Save Your Life
How To Inspect A Restaurant
Matches Aren't For Children
Public Relations for the Fire Service
Danger! Fire Fighters at Work, Safety 1 (NFPA)
The Fire Department Safety Program, Safety 2 (NFPA)**

TRANSPARENCIES

Multi-colored overhead transparencies to augment IFSTA 200, *Essentials of Fire Fighting*, are now available. Since costs and availability vary with different chapters, contact IFSTA Headquarters for details.

MANUAL BINDERS
Polyflex, expandable-pole binders hold 3 - 10 manuals.

GUIDE SHEET BINDERS
Free with purchase of basic 12 guide sets (101-110, 206, and Sprinklers).

WATER FLOW TEST SUMMARY SHEETS
50 summary sheets and instructions on how to use; logarithmic scale to simplify the process of determining the available water in an area.

PERSONNEL RECORD FOLDERS
Personnel record folders should be used by the training officer for each member of the department. Such data as IFSTA training, technical training (seminars), and college work can be recorded in this file, along with other valuable information. Letter size or legal size.